Selected Titles in This Series

745 **Mikhail A. Lifshits and Werner Linde,** Approximation and entropy numbers of Volterra operators with application to Brownian motion, 2002

744 **Roger Chalkley,** Basic global relative invariants for homogeneous linear differential equations, 2002

743 **Heng Sun,** Spectral decomposition of a covering of $GL(r)$: the Borel case, 2002

742 **J. E. Gilbert, Y. S. Han, J. A. Hogan, J. D. Lakey, D. Weiland, and G. Weiss,** Smooth molecular functions and singular integral operators, 2002

741 **Francisco Santos,** Triangulations of oriented matroids, 2002

740 **Rick Durrett,** Mutual invadability implies coexistence in spatial models, 2002

739 **Georgios K. Alexopoulos,** Sub-Laplacians with drift on Lie groups of polynomial volume growth, 2002

738 **Yasuro Gon,** Generalized Whittaker functions on $SU(2,2)$ with respect to the Siegel parabolic subgroup, 2002

737 **Arjen Doelman, Robert A. Gardner, and Tasso J. Kaper,** A stability index analysis of 1-D patterns of the Gray-Scott model, 2002

736 **Wojciech Chachólski and Jérôme Scherer,** Homotopy theory of diagrams, 2002

735 **Martina Brück, Xi Du, Joonsang Park, and Chuu-Lian Terng,** The submanifold geometries associated to Grassmannian systems, 2002

734 **Michel Van den Bergh,** Blowing up of non-commutative smooth surfaces, 2001

733 **Milé Krajčevski,** Tilings of the plane, hyperbolic groups and small cancellation conditions, 2001

732 **Jan O. Kleppe, Juan C. Migliore, Rosa Miró-Roig, Uwe Nagel, and Chris Peterson,** Gorenstein liaison, complete intersection liaison invariants and unobstructedness, 2001

731 **Jesús Bastero, Mario Milman, and Francisco J. Ruiz,** On the connection between weighted norm inequalities, commutators and real interpolation, 2001

730 **Suhyoung Choi,** The decomposition and classification of radiant affine 3-manifolds, 2001

729 **Michael Grosser, Eva Farkas, Michael Kunzinger, and Roland Steinbauer,** On the foundations of nonlinear generalized functions I and II, 2001

728 **Laura Smithies,** Equivariant analytic localization of group representations, 2001

727 **Anthony D. Blaom,** A geometric setting for Hamiltonian perturbation theory, 2001

726 **Victor L. Shapiro,** Singular quasilinearity and higher eigenvalues, 2001

725 **Jean-Pierre Rosay and Edgar Lee Stout,** Strong boundary values, analytic functionals, and nonlinear Paley-Wiener theory, 2001

724 **Lisa Carbone,** Non-uniform lattices on uniform trees, 2001

723 **Deborah M. King and John B. Strantzen,** Maximum entropy of cycles of even period, 2001

722 **Hernán Cendra, Jerrold E. Marsden, and Tudor S. Ratiu,** Lagrangian reduction by stages, 2001

721 **Ingrid C. Bauer,** Surfaces with $K^2 = 7$ and $p_g = 4$, 2001

720 **Palle E. T. Jorgensen,** Ruelle operators: Functions which are harmonic with respect to a transfer operator, 2001

719 **Steve Hofmann and John L. Lewis,** The Dirichlet problem for parabolic operators with singular drift terms, 2001

718 **Bernhard Lani-Wayda,** Wandering solutions of delay equations with sine-like feedback, 2001

717 **Ron Brown,** Frobenius groups and classical maximal orders, 2001

716 **John H. Palmieri,** Stable homotopy over the Steenrod algebra, 2001

(*Continued in the back of this publication*)

Approximation and Entropy Numbers of Volterra Operators with Application to Brownian Motion

Memoirs
of the
American Mathematical Society

Number 745

Approximation and Entropy
Numbers of Volterra Operators
with Application to
Brownian Motion

Mikhail A. Lifshits
Werner Linde

May 2002 • Volume 157 • Number 745 (first of 5 numbers) • ISSN 0065-9266

American Mathematical Society
Providence, Rhode Island

2000 *Mathematics Subject Classification.* Primary 47G10; Secondary 47B06, 60G15, 47B38.

Library of Congress Cataloging-in-Publication Data

Lifshits, M. A. (Mikhail Anatol′evich), 1956–
  Approximation and entropy numbers of Volterra operators with application to Brownian motion / Mikhail A. Lifshits, Werner Linde.
      p. cm. — (Memoirs of the American Mathematical Society, ISSN 0065-9266 ; no. 745)
  "Volume 157, number 745 (first of 5 numbers)."
  Includes bibliographical references and index.
  ISBN 0-8218-2791-X (alk. paper)
  1. Volterra operators. 2. Entropy (Information theory) 3. Brownian motion processes. I. Linde, Werner, 1947– . II. Title. III. Series.

QA3.A57 no. 745
[QA329.2]
510 s—dc21
[515′.723]   2002018235

Memoirs of the American Mathematical Society

This journal is devoted entirely to research in pure and applied mathematics.

Subscription information. The 2002 subscription begins with volume 155 and consists of six mailings, each containing one or more numbers. Subscription prices for 2002 are $524 list, $419 institutional member. A late charge of 10% of the subscription price will be imposed on orders received from nonmembers after January 1 of the subscription year. Subscribers outside the United States and India must pay a postage surcharge of $31; subscribers in India must pay a postage surcharge of $43. Expedited delivery to destinations in North America $35; elsewhere $130. Each number may be ordered separately; *please specify number* when ordering an individual number. For prices and titles of recently released numbers, see the New Publications sections of the *Notices of the American Mathematical Society*.

Back number information. For back issues see the *AMS Catalog of Publications*.

Subscriptions and orders should be addressed to the American Mathematical Society, P. O. Box 845904, Boston, MA 02284-5904. *All orders must be accompanied by payment.* Other correspondence should be addressed to Box 6248, Providence, RI 02940-6248.

Copying and reprinting. Individual readers of this publication, and nonprofit libraries acting for them, are permitted to make fair use of the material, such as to copy a chapter for use in teaching or research. Permission is granted to quote brief passages from this publication in reviews, provided the customary acknowledgment of the source is given.

Republication, systematic copying, or multiple reproduction of any material in this publication is permitted only under license from the American Mathematical Society. Requests for such permission should be addressed to the Acquisitions Department, American Mathematical Society, P. O. Box 6248, Providence, Rhode Island 02940-6248. Requests can also be made by e-mail to reprint-permission@ams.org.

Memoirs of the American Mathematical Society is published bimonthly (each volume consisting usually of more than one number) by the American Mathematical Society at 201 Charles Street, Providence, RI 02904-2294. Periodicals postage paid at Providence, RI. Postmaster: Send address changes to Memoirs, American Mathematical Society, P. O. Box 6248, Providence, RI 02940-6248.

© 2002 by the American Mathematical Society. All rights reserved.
This publication is indexed in *Science Citation Index*®, *SciSearch*®, *Research Alert*®, *CompuMath Citation Index*®, *Current Contents*®/*Physical, Chemical & Earth Sciences*.
Printed in the United States of America.

∞ The paper used in this book is acid-free and falls within the guidelines established to ensure permanence and durability.
Visit the AMS home page at URL: http://www.ams.org/

10 9 8 7 6 5 4 3 2 1   07 06 05 04 03 02

Contents

Chapter 1. Introduction	1
Chapter 2. Main Results	4
Chapter 3. Scale Transformations	9
3.1. Increasing Transformations	9
3.2. Decreasing Transformations	13
3.3. Examples	13
3.4. Transformations and Norms	17
Chapter 4. Upper Estimates for Entropy Numbers	20
4.1. A General Bound Based on Partitions	20
4.2. Proof of Theorem 2.2 (1)	22
4.3. Proof of Parts (2) and (3) in Theorem 2.2	25
4.4. Entropy Estimates for $T_{\rho,\psi}$	28
4.5. Proof of Theorem 2.3	32
4.6. Upper Bounds for Forward Integration Operators	37
4.7. Proof of Theorem 4.9	37
Chapter 5. Lower Estimates for Entropy Numbers	40
5.1. A General Construction for Lower Estimates	40
5.2. Proof of Theorem 2.4	43
5.3. Lower Bounds for Operators $T_{\rho,\psi}$ and $S_{\chi,\eta}$	46
5.4. Regular Kernel Functions	48
Chapter 6. Approximation Numbers	51
6.1. Maz'ja–Rosin Theorem	51
6.2. Estimates for $a_n(T_\rho)$ on Finite Intervals	53
6.3. Estimates for $a_n(T_\rho)$ in the Case $1 < p < 2 < q < \infty$	59
6.4. Approximation Numbers of T_ρ (General Case)	65
Chapter 7. Small Ball Behaviour of Weighted Wiener Processes	72
7.1. Gaussian Processes and Metric Entropy	72
7.2. Weighted Wiener Processes	74
7.3. Small Ball Estimates for Wiener Processes	76
7.4. Exact Small Ball Estimates	78
Appendix	85
Bibliography	86

Abstract

We consider the Volterra integral operator $T_{\rho,\psi} : L_p(0,\infty) \to L_q(0,\infty)$ for $1 \leq p, q \leq \infty$, defined by

$$(T_{\rho,\psi} f)(s) = \rho(s) \int_0^s \psi(t) f(t) dt$$

and investigate its degree of compactness in terms of properties of the kernel functions ρ and ψ. In particular, under certain optimal integrability conditions the entropy numbers $e_n(T_{\rho,\psi})$ satisfy

$$c_1 \|\rho \psi\|_r \leq \liminf_{n \to \infty} n \, e_n(T_{\rho,\psi}) \leq \limsup_{n \to \infty} n \, e_n(T_{\rho,\psi}) \leq c_2 \|\rho \psi\|_r$$

where $1/r = 1 - 1/p + 1/q > 0$. We also obtain similar sharp estimates for the approximation numbers of $T_{\rho,\psi}$, thus extending former results due to Edmunds et al. and Evans et al.. The entropy estimates are applied to investigate the small ball behaviour of weighted Wiener processes ρW in the $L_q(0,\infty)$–norm, $1 \leq q \leq \infty$. For example, if ρ satisfies some weak monotonicity conditions at zero and infinity, then

$$\lim_{\varepsilon \to 0} \varepsilon^2 \log \mathbb{P}(\|\rho W\|_q \leq \varepsilon) = -k_q \cdot \|\rho\|_{\frac{2q}{2+q}}^2 \ .$$

Received by the editor November 19, 1999.

1991 *Mathematics Subject Classification*. Primary: 47G10 ; Secondary: 47B06, 60G15, 47B38 .

Key words and phrases. Volterra operator, integral operator, entropy numbers, approximation numbers, Wiener process, small ball probabilities.

The first named author was supported in part by Grant INTAS 99-01317 and Grant RFBR 99-01-00112.

The second named author was supported in part by Grant RFBR/DFG 99-01-04027.

CHAPTER 1

Introduction

Linear integral operators in function spaces are among the most investigated objects in Functional Analysis, and their degree of compactness (depending on properties of the generating kernels) is of special interest. This is mainly due to the fact that this degree is tightly connected with the asymptotic behaviour of eigenvalues (cf. [10]). There exists a general very useful scheme for verifying compactness properties of integral operators. It is shown that certain properties of the kernel (smoothness, integrability etc.) yield a factorization of the generated operator via an embedding between suitable function spaces. Since in most cases the degree of compactness of those embeddings is well–known (cf. [19]), the integral operator possesses at least the same degree as the embedding. This method turned out to be very useful for a large class of operators (cf. [25] and [39]) and led to sharp estimates of their approximation numbers, entropy numbers or other quantities. Yet there exist several types of integral operators where this scheme does not apply. Among them is the important class of Volterra integral operators. Their investigation demands special treatment and techniques. For example, even in such a "simple" case as the ordinary d–dimensional integration operator T_d, $d \geq 3$, the exact behaviour of entropy numbers is still an open question (cf. [2], [43], [16] and [15]). In dimension one, it was an open problem for a long time to estimate the approximation numbers of $T_1 : L_p[0,1] \to L_q[0,1]$ for all pairs of p and q. The final solution (cf. Theorem 2.1 below) is based on a deep result of Gluskin (cf. [23]).

Motivated by probabilistic applications, we are interested in Volterra operators defined by

$$(1.1) \qquad T_{\rho,\psi} : f \longrightarrow \rho(s) \int_0^s \psi(t) f(t)\, dt$$

and mapping $L_p(0,\infty)$ into $L_q(0,\infty)$. The main concern here is the behavior of entropy numbers $e_n(T_{\rho,\psi})$. We were partially guided by results and methods of [18] (cf. also [20]), where approximation numbers of $T_{\rho,\psi}$ were investigated in the special case $p = q$. Our methods also apply to those numbers, so we were able to extend the results of [18] and [20] to almost all pairs of p and q.

The main estimate for entropy numbers asserts

$$(1.2) \qquad c_1 \|\rho\,\psi\|_r \leq \liminf_{n\to\infty} n e_n(T_{\rho,\psi}) \leq \limsup_{n\to\infty} n e_n(T_{\rho,\psi}) \leq c_2 \|\rho\,\psi\|_r$$

with $1/r = 1 - 1/p + 1/q > 0$, provided ρ and ψ satisfy certain integrability conditions. Although those conditions turn out to be the best possible for the validity of (1.2), they do not characterize functions ρ and ψ with $e_n(T_{\rho,\psi}) \approx n^{-1}$ completely. We feel that there exist some surprising phenomena not related with integrability but with some average continuity properties of the kernel.

Although our article is quite long, the basic ideas are rather easy to list:

- We use known approximation results for the well–investigated integration operator $T_{1,1}$ regarded on the interval $[0,1]$ (in what follows, we call it the *ordinary* integration operator).
- We use scale transformations which simplify our operators. In particular, we can often reduce the operator $T_{\rho,\psi}$ to an operator $T_{\tilde\rho,1}$ for suitable $\tilde\rho$. For the specific case $p=1$ we have to use a scale inversion jointly with important duality arguments due to Bourgain et al..
- We approximate the kernel functions by interval step functions.
- In some cases our arguments rely on deep results about approximation bounds for other types of operators, e.g. on Carl's approximation bounds for diagonal operators and on Gluskin's estimates for the approximation numbers of embeddings in finite dimensional l_p–spaces.
- Perhaps our main innovation is the "probabilistic" point of view to the problem. For example, the well–known decomposition of the Wiener process into a family of independent Brownian bridges and a piecewise linear process was the motive to introduce certain classes of subspaces in L_p (for the precise definition of the class $L_p^\circ(I)$ we refer to chapter 3, formula (3.31)). We use this technique many times when we want to reduce the study of integral operators to that of the ordinary integral operator.

The organization of the paper is as follows. The main estimates for $e_n(T_{\rho,\psi})$ are stated in chapter 2 and proved in chapter 4 (upper bounds) and chapter 5 (lower bounds), respectively. Moreover, we show that the integrability assumptions about ρ and ψ cannot be weakened. In the intermediate chapter 3, we develop the technique of variable changes. As one of the immediate corollaries of this technique we show how to reduce the investigation of general operators $T_{\rho,\psi}$ to that of the special case $T_\rho = T_{\rho,1}$ mapping $f \in L_p(0,\infty)$ onto $T_\rho f$ with

$$(1.3) \qquad (T_\rho f)(s) := \rho(s) \int_0^s f(t)\, dt, \quad s > 0.$$

Chapter 5 is devoted to lower estimates of $e_n(T_{\rho,\psi})$ which can be directly derived from a factorization of the ordinary integral operator by a multiple of T_ρ when suitably restricted to finite-codimensional subspaces.

In chapter 6 we apply our methods to the approximation numbers a_n of the operator $T_{\rho,\psi}$ acting from L_p into L_q. Here we extend earlier results of [17], [18] and [20] obtained in the case $1 \le p = q \le \infty$, to almost all pairs of p and q. Hence we get estimates for $n^\lambda a_n(T_{\rho,\psi})$ similarly to (1.2) with λ defined by (2.6) below.

Volterra integral operators are known to play an important role in the theory of stochastic processes. More precisely, given an orthonormal basis $(f_j)_{j=1}^\infty$ in $L_2(0,\infty)$ and an i.i.d. sequence $(\xi_j)_{j=1}^\infty$ of standard normal random variables, the stochastic process W defined by

$$(1.4) \qquad W(s) := \sum_{j=1}^\infty \xi_j \int_0^s f_j(t)\, dt, \quad s \ge 0,$$

is nothing else but the standard Wiener process. Hence, regarding T_ρ as operator from $L_2(0,\infty)$, in the same way as in (1.4) one obtains the weighted Wiener process

$$\rho(s)\, W(s) = \sum_{j=1}^\infty \xi_j (T_\rho f_j)(s), \quad s \ge 0.$$

There exists a tight connection between the small ball behaviour of a given Gaussian stochastic process and the entropy numbers of its generating operator (cf. [26] and [32]). Therefore our results on Volterra operators lead to quite sharp estimates for the probabilities

$$\mathbb{P}\left(\|\rho W\|_q < \varepsilon\right) = \mathbb{P}\left(\int_0^\infty |\rho(t)\,W(t)|^q\,dt < \varepsilon^q\right)$$

or, respectively,

$$\mathbb{P}\left(\sup_{t>0} |\rho(t)W(t)| < \varepsilon\right)$$

as $\varepsilon \to 0$. Using refined probabilistic methods we improve the general entropy based bounds for the small ball probabilities. We show the existence of so–called "small ball constants" for a large class of functions ρ. These results will be presented in chapter 7.

Acknowledgement: The authors are very grateful to H. Triebel for several helpful discussions, especially about Theorem 2.1. We express our thanks to D. E. Edmunds for drawing our attention to the article [20]. W.V. Li has kindly supplied us with recent preprints [29], [30] and [31]. The application of his new results helped us to simplify several proofs in the probabilistic part of the article. We also sincerely thank J. Creutzig for the careful reading of the manuscript and for his help to solve some TEX–problems. Finally, we express our thanks to J. Kuelbs who generously helped us to prepare the final version of the manuscript.

CHAPTER 2

Main Results

Given an operator T acting between Banach spaces E and F, its entropy and approximation numbers $e_n(T)$ and $a_n(T)$ are important quantities to measure the degree of compactness of T. They are defined in the following way (cf. [**10**] or [**40**]). Let $B_E = \{x \in E : \|x\| \leq 1\}$ be the unit ball in E and, if $y \in F$, then $B(y; \varepsilon)$ denotes the open ε–ball in F with center in y. With these notations

$$a_n(T) := \inf\{\|T - S\| : S : E \to F, \ \operatorname{rank}(S) < n\} \quad \text{and}$$

$$e_n(T) := \inf\left\{\varepsilon > 0 : T(B_E) \subseteq \bigcup_{j=1}^{2^{n-1}} B(Tx_j; \varepsilon) \text{ for some } x_1, \ldots, x_{2^{n-1}} \in B_E\right\}.$$

These quantities are related by the following important estimate (cf. [**7**] or [**9**] for an improved version). For any $\alpha > 0$ we have

$$(2.1) \qquad \sup_n n^\alpha e_n(T) \leq c_\alpha \sup_n n^\alpha a_n(T).$$

If $p, q \in [1, \infty]$, in all our further investigations the number $r > 0$ given by

$$(2.2) \qquad 1/r = 1/p' + 1/q$$

will be crucial. Here, as usual, p' defined by $1/p' = 1 - 1/p$ (with $\infty' = 1$ and $1' = \infty$) denotes the conjugate exponent of p. We always suppose $r < \infty$, i.e. we exclude the degenerated case $p = 1$ and $q = \infty$ where even on finite intervals the ordinary integral operator is not compact. Given p, q as before, functions denoted by ρ and ψ are always defined on $(0, \infty)$ with non–negative values such that

$$\psi \in L_p(0, x) \quad \text{and} \quad \rho \in L_q(x, \infty) \quad \text{for every } x > 0.$$

Then, if $f \in L_p(0, \infty)$, for all $s > 0$ the expression

$$(2.3) \qquad (T_{\rho,\psi} f)(s) := \rho(s) \int_0^s \psi(t) f(t)\, dt$$

is well–defined. Necessary and sufficient conditions for the boundedness of $T_{\rho,\psi}$ as operator from L_p into L_q will be stated and proved in Theorem 6.1 below.

If $\psi \equiv 1$, we write T_ρ instead of $T_{\rho,1}$ and T_1 will always denote the ordinary integral operator, i.e. $(T_1 f)(s) = \int_0^s f(t)\, dt$. As mentioned above, our main concern is the behaviour of $e_n(T_{\rho,\psi})$ as well as of $a_n(T_{\rho,\psi})$ as $n \to \infty$. Thus it is quite natural to first answer these questions for T_1 mapping $L_p(0,1)$ into $L_q(0,1)$. If $1 < p, q < \infty$, this can be found in [**19**], p. 118 and p. 119, respectively. In the remaining cases some additional considerations are necessary which will be carried out in an appendix. Unfortunately, the results in [**19**] do not provide the behaviour of $a_n(T_1 : L_p(0,1) \to L_q(0,1))$ for $p = 1$, $2 < q < \infty$ and in the dual case $q = \infty$, $1 < p < 2$.

THEOREM 2.1. *Regard T_1 as operator from $L_p(0,1)$ into $L_q(0,1)$. Then with $c_1, c_2 > 0$ only depending on p and q the following holds.*

(1) *If $1 \leq p, q \leq \infty$ and $1/p' + 1/q = 1/r > 0$, then*

(2.4) $$c_1 \, n^{-1} \leq e_n(T_1) \leq c_2 \, n^{-1} \, .$$

(2) *Let $1 \leq p, q \leq \infty$ with $1 \leq q \leq 2$ for $p = 1$ and $2 \leq p \leq \infty$ if $q = \infty$. Then we have*

(2.5) $$c_1 \, n^{-\lambda} \leq a_n(T_1) \leq c_2 \, n^{-\lambda}$$

where

(2.6) $$\lambda = \begin{cases} 1/r & : \ 1 \leq p \leq q \leq 2 \quad or \quad 2 \leq p \leq q \leq \infty \\ 1 & : \ 1 \leq q < p \leq \infty \\ 1/2 + \min\{1/q, 1/p'\} & : \ 1 \leq p \leq 2 \leq q \leq \infty \, . \end{cases}$$

The degree of compactness of T_ρ (later on of $T_{\rho,\psi}$) will be described by integrability conditions imposed on ρ or jointly on ρ and ψ, respectively. Besides the usual L_r–norm of ρ, i.e. $\|\rho\|_r = (\int_0^\infty \rho(t)^r \, dt)^{1/r}$, two other related quantities will play an important role. To introduce them let Δ_k, $k \in \mathbb{Z}$, be the dyadic intervals

(2.7) $$\Delta_k := [2^k, 2^{k+1}) \, .$$

For ρ as before, the numbers δ_k, $k \in \mathbb{Z}$, are defined by

(2.8) $$\delta_k = \delta_k(\rho) := 2^{k/p'} \|\rho\|_{L_q(\Delta_k)} = |\Delta_k|^{1/p'} \|\rho\|_{L_q(\Delta_k)} \, .$$

Then we set

(2.9) $$|\rho|_r := \|(\delta_k)_{k \in \mathbb{Z}}\|_r = \left(\sum_{k \in \mathbb{Z}} \delta_k^r \right)^{1/r}$$

and

(2.10) $$|\rho|_{r,\infty} := \sup_{k \geq 0}(k+1)^{1/r} \delta_k^* + \sup_{k \leq -1} |k|^{1/r} \delta_k^+$$

where $(\delta_k^*)_{k \geq 0}$ and $(\delta_k^+)_{k \leq -1}$ are the decreasing or increasing rearrangements of the δ_k's for $k \geq 0$ and $k \leq -1$, respectively. In other words,

$$|\rho|_{r,\infty} = \|(\delta_k)_{k \geq 0}\|_{r,\infty} + \|(\delta_k)_{k \leq -1}\|_{r,\infty}$$

with the usual Lorentz sequence norm $\| \cdot \|_{r,\infty}$ (cf. [**39**], section 2).

Hölder's inequality as well as inclusion properties of Lorentz sequence spaces yield

$$\|\rho\|_r \leq |\rho|_r \quad \text{and} \quad |\rho|_{r,\infty} \leq c \, |\rho|_r \, ,$$

respectively. On the other hand, there are easy examples of ρ's with $\|\rho\|_r < \infty$ and $|\rho|_{r,\infty} = \infty$ and vice versa, i.e. $\| \cdot \|_r$ and $| \cdot |_{r,\infty}$ are incomparable. Only under additional regularity conditions on ρ the r–integrability implies $|\rho|_r < \infty$ (cf. Corollary 2.5 below).

A special kernel regularization is needed for $q = \infty$ (cf. also [**20**]). Namely, let ρ be non–negative and assume that it belongs to $L_\infty(x, \infty)$ for each $x > 0$. The regularized function ρ^* is then defined by

(2.11) $$\rho^*(s) := \lim_{\delta \to 0} \operatorname*{ess\,sup}_{\{x : |x-s| < \delta\}} \rho(x) \, ,$$

and it is not difficult to see that

$$\|\rho \cdot g\|_\infty = \|\rho^* \cdot g\|_\infty$$

for any continuous g on $(0, \infty)$. Consequently, if T_ρ and T_{ρ^*} are regarded as operators with values in L_∞, for all $f \in L_p(0, \infty)$ necessarily

(2.12) $$\|T_\rho f\|_\infty = \|T_{\rho^*} f\|_\infty .$$

Since $L_\infty(0, \infty)$ has the metric extension property (cf. [38], C 3.2), both operators possess exactly the same degree of compactness. In particular, it follows that $e_n(T_\rho) = e_n(T_{\rho^*})$ and $a_n(T_\rho) = a_n(T_{\rho^*})$. Thus it is no surprise that for $q = \infty$, instead of norms of ρ, those of ρ^* appear in the estimates of $e_n(T_\rho)$ or $a_n(T_\rho)$, respectively. Observe that for very irregular ρ's it may be true that $\|\rho\|_r < \|\rho^*\|_r$, while by the definition we always have $\delta_k(\rho) = \delta_k(\rho^*)$, hence $|\rho|_r = |\rho^*|_r$. Similar phenomena appear by duality for operators $T_{\rho, \psi}$ acting on $L_1(0, \infty)$.

Now we are in the position to formulate the main results of the paper. For simplicity we restrict ourselves to the case $\psi \equiv 1$. For the proofs as well as for the results with general ψ's we refer to chapters 4–7 below. Here and in the sequel $c, C, \ldots$ are different constants which are allowed to depend on p, q but not on the kernel functions ρ, ψ and surely not on the dimension parameter n. Given two sequences a_n and b_n of positive real numbers we write $a_n \sim b_n$ provided that $a_n/b_n \to 1$ as $n \to \infty$ while $a_n \approx b_n$ means $c_1 a_n \leq b_n \leq c_2 a_n$ for $n \in \mathbb{N}$ and some $c_1, c_2 > 0$.

We start with upper estimates for the entropy numbers of T_ρ.

THEOREM 2.2. *For $1 \leq p, q \leq \infty$ let $r < \infty$ be defined by $1/r = 1/p' + 1/q$. Given ρ as before, the following estimates hold.*

(1) *For $T_\rho : L_p(0, \infty) \to L_q(0, \infty)$ we always have*

$$\sup_n n\, e_n(T_\rho) \leq c\, |\rho|_r .$$

(2) *Let $I \subseteq (0, \infty)$ be some bounded interval and suppose that $\rho \in L_q(I)$. Then regarding T_ρ as operator from $L_p(I)$ into $L_q(I)$ it follows that*

$$\limsup_{n \to \infty} n\, e_n(T_\rho) \leq c \cdot \|\rho\|_r$$

for $q < \infty$ while

$$\limsup_{n \to \infty} n\, e_n(T_\rho) \leq c \cdot \|\rho^*\|_{p'}$$

for $q = \infty$ with ρ^* defined by (2.11).

(3) *If $T_\rho : L_p(0, \infty) \to L_q(0, \infty)$, then $|\rho|_r < \infty$ also implies*

(2.13) $$\limsup_{n \to \infty} n\, e_n(T_\rho) \leq c \cdot \|\rho\|_r$$

with $\|\rho^*\|_{p'}$ on the right hand side of (2.13) for $q = \infty$.

The next result shows that the assumptions in Theorem 2.2 cannot be weakened.

THEOREM 2.3. *Suppose $p > 1$ and $1 \leq q \leq \infty$.*

(1) *For any non–negative q-summable sequence $(b_k)_{k=1}^\infty$ with $\sum_{k \geq 1} b_k^r = \infty$ we find a function ρ possessing the following properties:*
   (a) *The operator T_ρ is bounded from $L_p(0, \infty)$ to $L_q(0, \infty)$.*

(b) $\rho \in L_r(0,\infty)$ and also $\rho \in L_q(0,\infty)$.
(c) For each $k = 1, 2, \ldots$ we have $b_k = \delta_k(\rho) = 2^{k/p'} \|\rho\|_{L_q(\Delta_k)}$. In particular, $|\rho|_r = \infty$.
(d) $\limsup_{n \to \infty} n\, e_n(T_\rho) = \infty$.

(2) For every $M > 0$ and for each $n \in \mathbb{N}$ there exists a function $\rho \in L_r(0,1)$ (depending on n, M) such that $n\, e_n(T_\rho) \geq M \cdot \|\rho\|_r$.

The next result gives the corresponding lower estimates for entropy numbers of T_ρ.

THEOREM 2.4. *Suppose $1 \leq p, q \leq \infty$ and $1/r = 1/p' + 1/q > 0$. Let $\rho : (0,\infty) \to [0,\infty)$ be as before and regard T_ρ as operator from $L_p(0,\infty)$ into $L_q(0,\infty)$. For the entropy numbers of T_ρ the following lower estimates are valid:*

(1) *We have*

(2.14) $$|\rho|_{r,\infty} \leq c \cdot \sup_n n\, e_n(T_\rho)$$

with some universal $c > 0$.

(2) *It also is true that*

$$\|\rho\|_r \leq c \cdot \liminf_{n \to \infty} n\, e_n(T_\rho)$$

with $\|\rho^*\|_{p'}$ on the left hand side for $q = \infty$.

(3) *If $p > 1$, estimate (2.14) is optimal in the following sense: There exists a function ρ such that $e_n(T_\rho) \leq c\, n^{-1}$ and $\delta_k(\rho) = k^{-1/r}$, $k = 1, 2, \ldots$.*

Under special regularity assumptions about ρ the following stronger assertions are valid.

COROLLARY 2.5.
(1) Assume that for some real numbers α, β the functions $\rho(t)\, t^\alpha$ and $\rho(t)\, t^\beta$ are monotone in a neighborhood of zero and infinity, respectively. Then

$$c_1 \|\rho\|_r \leq \liminf_{n \to \infty} n\, e_n(T_\rho) \leq \limsup_{n \to \infty} n\, e_n(T_\rho) \leq c_2 \|\rho\|_r$$

with $\|\rho^*\|_{p'}$ on both sides for $q = \infty$.

(2) In the previous statement one may replace the assumption about monotonicity at zero by supposing $\rho \in L_q(0,\infty)$.

(3) Assume that for some real number α the function $\rho(t)\, t^\alpha$ is monotone on $(0, \infty)$. Then

$$c_1 \|\rho\|_r \leq \sup_n n\, e_n(T_\rho) \leq c_2\, c_\alpha \|\rho\|_r$$

with $c_\alpha = 2^{1/r + 2|\alpha|}$ and with $\|\rho^*\|_{p'}$ on both sides for $q = \infty$.

Remark: Essentially we do not need the monotonicity of ρ near zero and infinity. The weaker assumption

$$\limsup_{t \to \{0,\infty\}} \sup_{t' \in [t/2, 2t]} \rho(t')/\rho(t) < \infty$$

already suffices.

Our main result about approximation numbers asserts the following.

THEOREM 2.6. *Regard T_ρ as operator from $L_p(0,\infty)$ into $L_q(0,\infty)$ and define λ by (2.6).*

(1) *If either $1 \leq q < p \leq \infty$, $1 \leq p \leq q \leq 2$ or $2 \leq p \leq q \leq \infty$, then $|\rho|_r < \infty$ implies*

$$\limsup_{n\to\infty} n^\lambda a_n(T_\rho) \leq c \, \|\rho\|_r \tag{2.15}$$

with $\|\rho^*\|_{p'}$ on the right hand side of (2.15) for $q = \infty$.

(2) *If $1 < p < 2 < q < \infty$ and $|\rho|_{1/\lambda} = \left(\sum_{k\in\mathbb{Z}} \delta_k(\rho)^{1/\lambda}\right)^\lambda < \infty$, then*

$$\limsup_{n\to\infty} n^\lambda a_n(T_\rho) \leq c \cdot \inf_{\mathcal{I}} \left(\sum_{k\in\mathcal{K}} |I_k|^{1/(p'\lambda)} \|\rho\|_{L_q(I_k)}^{1/\lambda}\right)^\lambda \tag{2.16}$$

where the inf in (2.16) is taken over all countable partitions $\mathcal{I} = \{I_k\}_{k\in\mathcal{K}}$ of $(0,\infty)$ with

$$\text{card}\,\{k \in \mathcal{K} : I_k \cap A \neq \emptyset\} < \infty, \quad A \subset (0,\infty) \text{ compact.}$$

(3) *For all $p, q \in [1,\infty]$ as in (2) of Theorem 2.1 we have*

$$c \cdot \|\rho\|_r \leq \liminf_{n\to\infty} n^\lambda a_n(T_\rho)$$

with $\|\rho^*\|_{p'}$ on the left-hand side for $q = \infty$. Especially, under the assumptions (1) or (2) of Corollary 2.5 and for p and q as in (2.15) it follows that

$$c_1 \cdot \|\rho\|_r \leq \liminf_{n\to\infty} n^\lambda a_n(T_\rho) \leq \limsup_{n\to\infty} n^\lambda a_n(T_\rho) \leq c_2 \, \|\rho\|_r$$

with concrete constants $c_1, c_2 > 0$ depending on p and q and $\|\rho^*\|_{p'}$ on both sides for $q = \infty$.

The analytic results have interesting consequences for the small ball behavior of weighted Wiener processes. Let $W(t)$, $0 < t < \infty$, denote the classical Wiener process.

THEOREM 2.7. *Let $1 \leq q \leq \infty$ and define $r < \infty$ by $1/r = 1/2 + 1/q$. If $|\rho|_r < \infty$, then*

$$\lim_{\varepsilon\to 0} \varepsilon^2 \log \mathbb{P}(\|\rho W\|_q \leq \varepsilon) = -k_q \cdot \|\rho\|_r^2, \quad q < \infty, \quad \text{and}$$

$$\lim_{\varepsilon\to 0} \varepsilon^2 \log \mathbb{P}(\|\rho W\|_\infty \leq \varepsilon) = -(\pi^2/8) \cdot \|\rho^*\|_2^2 \,.$$

Here

$$k_q := -\lim_{\varepsilon\to 0} \varepsilon^2 \log \mathbb{P}\left(\|W\|_{L_q[0,1]} \leq \varepsilon\right).$$

In particular, we have

COROLLARY 2.8. *Suppose that $\rho \in L_r(0,\infty)$ satisfies the assumptions of Corollary 2.5. Then*

$$\lim_{\varepsilon\to 0} \varepsilon^2 \log \mathbb{P}(\rho \|W\|_q \leq \varepsilon) = -k_q \cdot \|\rho\|_r^2 \,.$$

This Corollary was recently proved by W. V. Li in [29], [30] for Riemann–integrable functions ρ under slightly stronger regularity assumptions at zero and infinity.

CHAPTER 3

Scale Transformations

Our objective is to show how variable changes may considerably simplify the structure of $T_{\rho,\psi}$. For example, if $\psi > 0$ a.e., by a suitable transformation we may replace ψ by $\tilde{\psi} \equiv 1$, so that in most cases we have to deal with one kernel function only. Using another transformation, we show that the dual operator $T'_{\rho,\psi}$ possesses the same structure as $T_{\rho,\psi}$ itself. Finally, piecewise linear scale transformations are crucial in our further investigations. All these different transformations may be treated uniformly; we only have to distinguish between increasing and decreasing transformations. Let us start with increasing ones.

3.1. Increasing Transformations

Consider the function $\theta : (0, \infty) \to (0, \infty)$ of the form

$$(3.1) \qquad \theta(s) = \int_0^s \theta'(t)\, dt, \quad s > 0,$$

for some $\theta' \in L_1(0, x)$, $x > 0$, with $\theta' \geq 0$. If necessary, by changing θ' on a set of measure zero, we may always suppose that $\theta'(t)$ coincides with the derivative of θ in t provided it exists. Let $I \subseteq (0, \infty)$ be some bounded or unbounded interval, then $A = \theta(I)$ denotes the image of I under the transformation θ. Since we do not assume $\theta' > 0$ a.e., the mapping θ need not to be invertible. Hence the left pseudoinverse

$$\theta^-(\tau) := \inf\{t \in I : \theta(t) = \tau\}, \quad \tau \in A,$$

has to serve as a substitute for θ^{-1}. We state now some well-known properties of θ^- for later application:
(1) θ^- is continuous from the left and increasing.
(2) For each $\tau \in A$ we have $\theta(\theta^- \tau) = \tau$, yet in general we only have $\theta^-(\theta t) \leq t$ for $t \in I$.
(3) Let λ be the Lebesgue measure on A and define a measure ν on I by

$$\nu(B) := \int_B \theta'(t)\, dt$$

for measurable $B \subseteq I$. Then

$$(3.2) \qquad \nu = \lambda \circ (\theta^-)^{-1},$$

i.e. ν is the image of λ under θ^-. In the language of integrals this may be written as

$$(3.3) \qquad \int_A f(\theta^- \tau)\, d\tau = \int_I f(t)\, \theta'(t)\, dt$$

for each $f \in L_1(I,\nu)$. Applying this to $f = g \circ \theta$ for a measurable function $g: A \to R$, we have the well–known transformation formula

$$(3.4) \qquad \int_A g(\tau)\,d\tau = \int_I g(\theta t)\,\theta'(t)\,dt$$

provided the right hand integral exists.

(4) If

$$N_\theta := \{t \in I : \theta'(t) > 0\}\ ,$$

then by the choice of θ' it follows

$$(3.5) \qquad \theta^-(\theta t) = t \quad \text{for all} \quad t \in N_\theta\ .$$

Moreover, we claim that

$$(3.6) \qquad \theta(N_\theta) = (\theta^-)^{-1}(N_\theta)$$

and

$$(3.7) \qquad \lambda\{A \setminus \theta(N_\theta)\} = 0.$$

Indeed, if $\tau \in \theta(N_\theta)$, then there exists a $t \in N_\theta$ with $\tau = \theta(t)$, hence (3.5) implies

$$\theta^-\tau = \theta^-\theta t = t$$

and $\tau \in (\theta^-)^{-1}(N_\theta)$. Conversely, if $\tau \in (\theta^-)^{-1}(N_\theta)$, then $t = \theta^-\tau \in N_\theta$, hence (3.5) ensures $\theta t = \theta\theta^-\tau = \tau$, which means $\tau \in \theta(N_\theta)$ and (3.6) is valid. To verify (3.7) observe that

$$A \setminus \theta(N_\theta) \subset (\theta^-)^{-1}(I \setminus N_\theta)$$

and applying (3.3) with $f = \mathbf{1}_{I \setminus N_\theta}$ yields

$$\int_{(\theta^-)^{-1}(I \setminus N_\theta)} d\tau = \int_{I \setminus N_\theta} \theta'(t)\,dt = 0$$

and (3.7) follows.

(5) If g_1 and g_2 on A are Lebesgue–equivalent, i.e.

$$\lambda\{\tau \in A :\ g_1(\tau) \neq g_2(\tau)\} = 0,$$

by (3.2) we have

$$\lambda\{t \in N_\theta : g_1(\theta t) \neq g_2(\theta t)\} = 0.$$

In other words, we may regard g in the transformation formula (3.4) as an element of $L_1(A, d\tau)$ and write

$$(3.8) \qquad \int_A g(\tau)\,d\tau = \int_{N_\theta} g(\theta t)\,\theta'(t)\,dt\ .$$

For example, if $\alpha, \beta \in A$ with $\alpha < \beta$ we may replace in (3.8) the function g by $g \cdot \mathbf{1}_{[\alpha,\beta)}$, and using (3.5) we obtain

$$(3.9) \qquad \int_\alpha^\beta g(\tau)\,d\tau = \int_{\theta^{-1}([\alpha,\beta)) \cap N_\theta} g(\theta t)\,\theta'(t)\,dt = \int_{\theta^-\alpha}^{\theta^-\beta} g(\theta t)\,\theta'(t)\,dt\ .$$

A change of the scale parameter generates a natural transformation of functions depending on this parameter. Namely, given $p \in [1,\infty]$ and a function $g \in L_1(A, d\tau)$, we define the corresponding transformed function, for $t \in N_\theta$, by

$$(3.10) \qquad (\Phi_p^\theta g)(t) := g(\theta t)\,\theta'(t)^{1/p} \quad \text{for} \quad p < \infty \quad \text{and}$$

$$(3.11) \qquad (\Phi_\infty^\theta g)(t) := g(\theta t) \quad \text{for} \quad p = \infty\ .$$

In view of (3.8) the following is valid.

PROPOSITION 3.1. *For $1 \leq p \leq \infty$ the mapping Φ_p^θ is an isometry between $L_p(A, d\tau)$ and $L_p(N_\theta, dt)$.*

We may and do identify $L_p(N_\theta, dt)$ with the subspace E_p^θ of $L_p(I, dt)$ defined by

(3.12) $$E_p^\theta := \{f \in L_p(I, dt) : f = f \cdot \mathbf{1}_{N_\theta} \text{ a.e.}\} .$$

In this interpretation, Φ_p^θ is an isometric embedding of $L_p(A, d\tau)$ into $L_p(I, dt)$ with image E_p^θ. If $\theta' > 0$ a.e. (but not in general), Φ_p^θ turns out to be an isometry *onto* $L_p(I, dt)$.

For $f \in E_p^\theta$ we define the inverse operator by

$$((\Phi_p^\theta)^{-1} f)(\tau) := f(\theta^- \tau) \cdot \theta'(\theta^- \tau)^{-1/p}, \quad \tau \in A.$$

In fact, by (3.6) and (3.7), this expression is well defined for almost all $\tau \in A$ and we obviously have

$$((\Phi_p^\theta)^{-1} \Phi_p^\theta g)(\tau) = g(\theta \theta^- \tau) \, \theta'(\theta^- \tau)^{1/p} \, \theta'(\theta^- \tau)^{-1/p} = g(\tau).$$

We consider now the influence of the scale transformation on the operator $T_{\rho,\psi}$. In other words, one asks for a Volterra operator $T_{\tilde{\rho}, \tilde{\psi}} : L_p(A) \to L_q(A)$ such that

$$T_{\tilde{\rho}, \tilde{\psi}} = (\Phi_q^\theta)^{-1} \circ T_{\rho, \psi} \circ \Phi_p^\theta .$$

More precisely, let

(3.13) $$\bar{\rho} := \rho \cdot \mathbf{1}_{N_\theta} \quad \text{and} \quad \bar{\psi} := \psi \cdot \mathbf{1}_{N_\theta}$$

be the restrictions of ρ and ψ on N_θ, respectively. Then $T_{\bar{\rho}, \bar{\psi}}$ denotes the corresponding restriction of the Volterra operator with kernel functions $\bar{\rho}, \bar{\psi}$ acting from E_p^θ into E_q^θ. Furthermore, we define the transformed kernel functions by

(3.14) $$\tilde{\rho}(\tau) := \rho(\theta^- \tau) \cdot \theta'(\theta^- \tau)^{-1/q}, \quad \tau \in A, \quad \text{and}$$

(3.15) $$\tilde{\psi}(\tau) := \psi(\theta^- \tau) \cdot \theta'(\theta^- \tau)^{-1/p'}, \quad \tau \in A,$$

with necessary modification as in (3.11) for $q = \infty$ or $p = 1$, respectively. Consider now the Volterra operator $T_{\tilde{\rho}, \tilde{\psi}} : L_p(A) \to L_q(A)$ defined by

(3.16) $$(T_{\tilde{\rho}, \tilde{\psi}} g)(\sigma) = \tilde{\rho}(\sigma) \int_0^\sigma \tilde{\psi}(\tau) \, g(\tau) \, d\tau, \quad \sigma \in A.$$

With these notations the following is valid.

PROPOSITION 3.2. *Let θ on I be defined by (3.1) with range $A = \theta(I)$. For the operators $T_{\bar{\rho}, \bar{\psi}}$ and $T_{\tilde{\rho}, \tilde{\psi}}$ defined as above, the following relations hold.*

(3.17) $$T_{\bar{\rho}, \bar{\psi}} = \Phi_q^\theta \circ T_{\tilde{\rho}, \tilde{\psi}} \circ (\Phi_p^\theta)^{-1} \quad \text{and}$$

(3.18) $$T_{\tilde{\rho}, \tilde{\psi}} = (\Phi_q^\theta)^{-1} \circ T_{\bar{\rho}, \bar{\psi}} \circ \Phi_p^\theta .$$

Furthermore, if $\theta' > 0$ a.e., then (3.17) and (3.18) are valid for $T_{\rho,\psi}$ instead of $T_{\bar{\rho}, \bar{\psi}}$.

PROOF. Given $g \in L_p(A)$, it follows from (3.10), (3.13), (3.5), (3.4) and (3.15) that

$$
\begin{aligned}
(T_{\bar{\rho},\bar{\psi}} \circ \Phi_p^\theta g)(s) &= \bar{\rho}(s) \int_0^s \bar{\psi}(t)\, g(\theta t)\, \theta'(t)^{1/p}\, dt \\
&= \bar{\rho}(s) \int_{I \cap N_\theta} \mathbf{1}_{[0,s]}(t)\, \psi(t)\, g(\theta t)\, \theta'(t)^{1/p}\, dt \\
&= \bar{\rho}(s) \int_I \psi(\theta^- \theta t)\, \mathbf{1}_{[0,s]}(\theta^- \theta t)\, g(\theta t)\, \theta'(\theta^- \theta t)^{1/p}\, dt \\
&= \bar{\rho}(s) \int_A \psi(\theta^- \tau)\, \mathbf{1}_{[0,s]}(\theta^- \tau)\, g(\tau)\, \theta'(\theta^- \tau)^{-1/p'}\, d\tau \\
&= \bar{\rho}(s) \int_{\theta^- \tau \leq s} \tilde{\psi}(\tau)\, g(\tau)\, d\tau \, .
\end{aligned}
$$

On the other hand, for $s \in N_\theta$ we have by (3.10), (3.16), (3.14) and (3.5)

$$
\begin{aligned}
(\Phi_q^\theta \circ T_{\tilde{\rho},\tilde{\psi}} g)(s) &= (T_{\tilde{\rho},\tilde{\psi}} g)(\theta s) \cdot \theta'(s)^{1/q} \\
&= \tilde{\rho}(\theta s)\, \theta'(s)^{1/q} \int_0^{\theta s} \tilde{\psi}(\tau)\, g(\tau)\, d\tau \\
&= \rho(\theta^- \theta s)\, \theta'(\theta^- \theta s)^{-1/q}\, \theta'(s)^{1/q} \int_0^{\theta s} \tilde{\psi}(\tau)\, g(\tau)\, d\tau \\
&= \bar{\rho}(s) \int_0^{\theta s} \tilde{\psi}(\tau)\, g(\tau)\, d\tau \\
&= \bar{\rho}(s) \int_{\tau \leq \theta s} \tilde{\psi}(\tau)\, g(\tau)\, d\tau \, .
\end{aligned}
$$

Note that $\theta^- \tau \leq s$ implies $\theta \theta^- \tau \leq \theta s$, hence $\tau \leq \theta s$. On the other hand $\tau \leq \theta s$ implies $\theta^- \tau \leq \theta^- \theta s \leq s$, so that $\tau \leq \theta s$ iff $\theta^- \tau \leq s$, and this observation yields

$$T_{\bar{\rho},\bar{\psi}} \circ \Phi_p^\theta = \Phi_q^\theta \circ T_{\tilde{\rho},\tilde{\psi}}.$$

Now (3.17) and (3.18) follow by composing with $(\Phi_p^\theta)^{-1}$ and $(\Phi_q^\theta)^{-1}$, respectively. □

COROLLARY 3.3. We always have

$$e_n(T_{\tilde{\rho},\tilde{\psi}}) = e_n(T_{\bar{\rho},\bar{\psi}}) \leq e_n(T_{\rho,\psi})$$

with equality in the case $\theta' > 0$ a.e..

PROOF. Note that $T_{\bar{\rho},\bar{\psi}} = M_q^\theta \circ T_{\rho,\psi} \circ M_p^\theta$ where M_p^θ denotes the multiplication operator in $L_p(I, dt)$ generated by $\mathbf{1}_{N_\theta}$. Since $\left\| M_p^\theta : L_p(I, dt) \to L_p(I, dt) \right\| \leq 1$, we have $e_n(T_{\bar{\rho},\bar{\psi}}) \leq e_n(T_{\rho,\psi})$ by well-known properties of the entropy numbers. □

Remark: Let us say that two operators T and S are isomorphic provided that there are suitable isometric operators $J_1, J_2, \hat{J}_1$ and $\hat{J}_2$ such that $T = J_1 \circ S \circ J_2$ and $S = \hat{J}_2 \circ T \circ \hat{J}_1$. In this notation Proposition 3.2 asserts that $T_{\rho,\psi}$ and $T_{\tilde{\rho},\tilde{\psi}}$ are isomorphic provided that $\theta' > 0$ a.e. on I.

3.2. Decreasing Transformations

Later on we also need decreasing transformations, i.e. consider

$$(3.19) \qquad \theta(s) = \int_s^\infty |\theta'(t)|\, dt\,, \quad s > 0,$$

for some $\theta' \leq 0$ with $\theta' \in L_1(x,\infty)$ for each $x > 0$. For $I \subseteq (0,\infty)$ let A again be the range of θ, i.e. θ transforms I onto A. Almost all properties of θ, as e.g. (3.2), (3.4) or Proposition 3.1, remain valid. However, notice the following changes.
(1) The pseudoinverse θ^- now needs to be defined as

$$\theta^-(\tau) := \sup\{t \in I : \theta(t) = \tau\}\,.$$

(2) In general we only have $\theta^-(\theta t) \geq t$, but equality holds for $t \in N_\theta$ defined this time by

$$N_\theta := \{t \in I : |\theta'(t)| > 0\}\,.$$

(3) In the definition of the transfer operators Φ_p^θ (see (3.10)), we have to replace θ' by $|\theta'|$.
(4) There is an important modification in Proposition 3.2. To state it, we need forward integration operators $S_{\chi,\eta} : L_p(0,\infty) \to L_q(0,\infty)$. Let us suppose that χ, η are functions from $(0,\infty)$ to $[0,\infty)$ with

$$(3.20) \qquad \chi \in L_q(0,\sigma) \quad \text{and} \quad \eta \in L_{p'}(\sigma,\infty) \quad \text{for every } \sigma > 0\,.$$

Then for $g \in L_p(0,\infty)$ we define

$$(3.21) \qquad (S_{\chi,\eta} g)(\sigma) := \chi(\sigma) \int_\sigma^\infty \eta(\tau) g(\tau)\, d\tau\,, \quad \sigma > 0\,.$$

With this notation the following holds.

PROPOSITION 3.4. *For decreasing scale transformations θ defined in (3.19), we have*

$$\begin{aligned} T_{\bar{\rho},\bar{\psi}} &= \Phi_q^\theta \circ S_{\tilde{\rho},\tilde{\psi}} \circ (\Phi_p^\theta)^{-1} \quad \text{and} \\ S_{\tilde{\rho},\tilde{\psi}} &= (\Phi_q^\theta)^{-1} \circ T_{\bar{\rho},\bar{\psi}} \circ \Phi_p^\theta \end{aligned}$$

where $\bar{\rho}, \bar{\psi}, \tilde{\rho}, \tilde{\psi}$, are defined by (3.13), (3.14) and (3.15), respectively.

PROOF. We proceed exactly as in the proof of Proposition 3.2. The only difference is that no longer $\tau \leq \theta s$ iff $\theta^- \tau \leq s$. Instead we have now $\tau \geq \theta s$ iff $\theta^- \tau \leq s$ leading to the change from $T_{\bar{\rho},\bar{\psi}}$ to $S_{\tilde{\rho},\tilde{\psi}}$ and this completes the proof. $\square$

COROLLARY 3.5. *We always have*

$$e_n(S_{\tilde{\rho},\tilde{\psi}}) \leq e_n(T_{\rho,\psi})$$

with equality when $|\theta'| > 0$ a.e. on I.

3.3. Examples

Now we present the main scale transformations used later on:
(a) For $p > 1$ and ρ and ψ as before let

$$\theta'(t) := \psi(t)^{p'}, \quad t \in (0,\infty)\,.$$

Then we have

(3.22) $$\theta(s) = \int_0^s \psi(t)^{p'} \, dt, \quad s > 0,$$

with $I = (0, \infty)$ and range $A = (0, \theta(\infty))$ where $\theta(\infty) = \|\psi\|_{p'}^{p'}$ may be infinite. The pseudoinverse $\theta^- : A \to (0, \infty)$ is thus given by

(3.23) $$\theta^-(\tau) = \inf \left\{ s > 0 : \int_0^s \psi(t)^{p'} \, dt = \tau \right\}$$

and, of course, if $\psi > 0$ a.e., then $\theta^- = \theta^{-1}$. The transformed functions $\tilde{\rho}$ and $\tilde{\psi}$ (according to (3.14) and (3.15), respectively) are in this case given by

(3.24) $$\begin{aligned} \tilde{\rho}(\tau) &= \rho(\theta^-\tau) \cdot \psi(\theta^-\tau)^{-p'/q}, \quad \tau \in A, \quad \text{and} \\ \tilde{\psi}(\tau) &\equiv 1. \end{aligned}$$

So Proposition 3.2 reads here as follows.

PROPOSITION 3.6. *Suppose $\rho = \rho \cdot \mathbf{1}_{\{\psi > 0\}}$ a.e. on $(0, \infty)$. Then for $p > 1$ and $\tilde{\rho}$ defined by (3.24) the operators $T_{\rho,\psi}$ and $T_{\tilde{\rho}}$ are isomorphic. In particular, if $\psi > 0$ a.e., then by changing ρ suitably, ψ can always be replaced by $\tilde{\psi} \equiv 1$.*

(b) Next we want to apply the general construction to a scale inversion. Let θ on $(0, \infty)$ be defined by

(3.25) $$\theta(s) = s^{-1} = \int_s^\infty t^{-2} \, dt, \quad s > 0.$$

In this case the transformed functions are (with obvious modifications for $p = 1$ and $q = \infty$)

(3.26) $$\tilde{\rho}(\tau) = \rho(1/\tau) \cdot \tau^{-2/q} \quad \text{and}$$
(3.27) $$\tilde{\psi}(\tau) = \psi(1/\tau) \cdot \tau^{-2/p'}$$

and the following special case of Proposition 3.4 is valid.

PROPOSITION 3.7. *The operator $T_{\rho,\psi}$ is isomorphic to $S_{\tilde{\rho},\tilde{\psi}} : L_p(0, \infty) \to L_q(0, \infty)$. Conversely, if χ, η satisfy (3.20), then $S_{\chi,\eta}$ is isomorphic to $T_{\tilde{\chi},\tilde{\eta}}$ with $\tilde{\chi}$ and $\tilde{\eta}$ defined by*

$$\tilde{\chi}(t) = \chi(1/t) \cdot t^{-2/q}$$

and

$$\tilde{\eta}(t) = \eta(1/t) \cdot t^{-2/p'},$$

respectively.

Let us give here a first immediate application. If $T_{\rho,\psi}$ maps $L_p(0, \infty)$ to $L_q(0, \infty)$, by $T'_{\rho,\psi}$ from $L_{q'}(0, \infty)$ to $L_{p'}(0, \infty)$ we denote its dual operator. Direct calculations show $T'_{\rho,\psi} = S_{\psi,\rho}$. Hence Proposition 3.7 yields the following useful result.

PROPOSITION 3.8. *For each pair (ρ, ψ) the dual operator $T'_{\rho,\psi}$ is isomorphic to $T_{\tilde{\psi},\tilde{\rho}} : L_{q'} \to L_{p'}$ with $\tilde{\psi}$ and $\tilde{\rho}$ defined by (3.27) and (3.26), respectively.*

PROOF. This follows directly from Proposition 3.7. Note that we have to replace ρ by ψ and vice versa. Moreover, the number p becomes q' and q has to be replaced by p'. □

(c) If $q < \infty$ we now define a (decreasing) scale transformation by

$$\theta(s) := \int_s^\infty \rho(t)^q \, dt, \quad s > 0, \tag{3.28}$$

hence

$$\theta^-(\tau) = \sup\left\{s > 0 : \int_s^\infty \rho(t)^q \, dt = \tau\right\}.$$

The image of θ is the interval $(0, \theta(0))$ with $\theta(0) = \|\rho\|_q^q$ which may be infinite. The transformation (3.28) leads to the functions

$$\tilde{\rho}(\tau) \equiv 1 \quad \text{and}$$
$$\tilde{\psi}(\tau) = \psi(\theta^-\tau) \cdot \rho(\theta^-\tau)^{-q/p'}, \quad 0 < \tau < \theta(0).$$

Hence from Proposition 3.4 we derive the following.

PROPOSITION 3.9. *For $q < \infty$ and $\psi = \psi \cdot \mathbf{1}_{\{\rho > 0\}}$ a.e. the operator $T_{\rho,\psi} : L_p(0, \infty) \to L_q(0, \infty)$ is isomorphic to $S_{1,\tilde{\psi}}$ mapping $L_p(0, \theta(0))$ into the corresponding L_q-space.*

Remark: Recall that

$$(S_{1,\tilde{\psi}} g)(\sigma) = \int_\sigma^\infty \tilde{\psi}(\tau) g(\tau) \, d\tau.$$

For $p = 1$ and $\psi \equiv 1$ the situation is even easier. Here we have $\tilde{\psi} \equiv 1$ as well, and if $\rho > 0$ a.e., then T_ρ as an operator from L_1 into L_q is isomorphic to S_1 defined on the interval $(0, \|\rho\|_q^q)$. Yet the operator S_1 is bounded if and only if the interval is finite, i.e. we necessarily have $\|\rho\|_q < \infty$. Moreover, we may apply a linear inversion of the interval $(0, \|\rho\|_q^q)$, namely,

$$\theta(t) = \|\rho\|_q^q - t,$$

and, using Proposition 3.4, we observe that S_1 is isomorphic to T_1. Consequently we get the following.

PROPOSITION 3.10. *If $p = 1$, $\rho > 0$ a.e. and $T_\rho : L_1(0, \infty) \to L_q(0, \infty)$ is bounded, then T_ρ is isomorphic to T_1 over a finite interval of length $\|\rho\|_q^q$.*

(d) Define

$$\theta'(t) = \mathbf{1}_G(t), \quad t > 0,$$

for some measurable subset $G \subseteq (0, \infty)$. Then

$$\theta(s) = |G \cap (0, s]|, \quad s > 0,$$

and the image of the transformation coincides with the interval $A = (0, |G|)$. Given ρ and ψ as before, their restrictions are now defined by

$$\bar{\rho} = \rho \cdot \mathbf{1}_G \quad \text{and} \quad \bar{\psi} = \psi \cdot \mathbf{1}_G,$$

and

$$(T_{\bar{\rho},\bar{\psi}} f)(s) = \mathbf{1}_G(s) \rho(s) \int_0^s f(t) \psi(t) \mathbf{1}_G(t) \, dt.$$

Therfore $T_{\bar{\rho},\bar{\psi}}$ is isomorphic to $T_{\tilde{\rho},\tilde{\psi}}$ acting on $(0, |G|)$. Since (3.2) implies

$$\lambda\{\tau : \theta^-\tau \in G, \theta'(\theta^-\tau) = 1\} = \lambda\{t \in G : \theta'(t) = 1\} = |G|,$$

we have $\mathbf{1}_G(\theta^-\tau) = 1$ a.e. on A, hence the transformed functions $\tilde\rho$ and $\tilde\psi$ coincide with
$$\tilde\rho(\tau) = \rho(\theta^-\tau) \cdot \mathbf{1}_G(\theta^-\tau)^{-1/q} = \rho(\theta^-\tau) \quad \text{and}$$
$$\tilde\psi(\tau) = \psi(\theta^-\tau) \cdot \mathbf{1}_G(\theta^-\tau)^{-1/p'} = \psi(\theta^-\tau), \quad \tau \in (0, |G|).$$
In particular, setting $G = \{\rho > 0\}$, it follows $\tilde\rho(\tau) > 0$ a.e. on $(0, |G|)$, i.e. we may transform $T_{\rho,\psi}$ to an operator $T_{\tilde\rho,\tilde\psi}$ with $\tilde\rho > 0$ at the cost of a projection $f \to f \cdot \mathbf{1}_G$ on $L_p(0, \infty)$. For example, if $\psi \equiv 1$, then for G as before $T_{\rho, \mathbf{1}_G}$ changes to $T_{\tilde\rho}$ on $(0, |G|)$ with $\tilde\rho > 0$ a.e.. Hence, any ρ may be transformed in this way to some $\tilde\rho > 0$ with
$$e_n(T_{\tilde\rho}) = e_n(T_{\rho, \mathbf{1}_G}) \le e_n(T_\rho).$$

(e) Let $I \subseteq (0, \infty)$ be some interval and suppose that $I = \bigcup_{k \in \mathcal{K}} I_k$ for some family of disjoint finite intervals I_k. Given $\lambda_k > 0$ we now set

(3.29) $$\theta'(t) := \lambda_k, \quad t \in I_k,$$

and hence θ is a piecewise linear transformation from I onto $A = \theta(I)$. If θ_k denotes the restriction of θ onto I_k, and $A_k := \theta_k(I_k)$ then $A = \bigcup_{k \in \mathcal{K}} A_k$ with A_k's being disjoint intervals. Note that
$$|A_k| = \lambda_k \cdot |I_k|, \quad k \in \mathcal{K}.$$
The isometry Φ_p^θ from $L_p(A)$ onto $L_p(I)$ is then given by
$$(\Phi_p^\theta g)(t) = \lambda_k^{1/p} \cdot g(\theta_k t), \quad t \in I_k,$$
while
$$((\Phi_p^\theta)^{-1} f)(\tau) = \lambda_k^{-1/p} \cdot f(\theta_k^{-1} \tau), \quad \tau \in A_k.$$
The transformed ρ is now $\tilde\rho$ on A defined by
$$\tilde\rho(\tau) = \lambda_k^{-1/q} \cdot \rho(\theta_k^{-1} \tau), \quad \tau \in A_k,$$
and for $\psi \equiv 1$ we get
$$\tilde\psi(\tau) = \lambda_k^{-1/p'}, \quad \tau \in A_k.$$
Consequently, the operator $T_\rho : L_p(I) \to L_q(I)$ is isomorphic to $T_{\tilde\rho, \tilde\psi}$ on $L_p(A)$ with

(3.30) $$(T_{\tilde\rho, \tilde\psi} g)(\sigma) = \lambda_k^{-1/q} \rho(\theta_k^{-1} \sigma) \int_0^\sigma g(\tau) \tilde\psi(\tau) \, d\tau, \quad \sigma \in A_k.$$

For special g's in $L_p(A)$ expression (3.30) becomes much simpler. Here we introduce subspaces $L_p^\circ(I)$ and $L_p^\circ(A)$ as follows:

(3.31) $$L_p^\circ(I) := \left\{ f \in L_p(I) : \int_{I_k} f(t) \, dt = 0, \ k \in \mathcal{K} \right\}$$

and similarly,
$$L_p^\circ(A) := \left\{ g \in L_p(A) : \int_{A_k} g(\tau) \, d\tau = 0, \ k \in \mathcal{K} \right\}.$$
Note that Φ_p^θ maps $L_p^\circ(A)$ isometrically onto $L_p^\circ(I)$, so if T_ρ° denotes the restriction of T_ρ onto $L_p^\circ(I)$, it follows that

(3.32) $$T_\rho^\circ = \Phi_q^\theta \circ T_{\tilde\rho, \tilde\psi}^\circ \circ (\Phi_p^\theta)^{-1}$$

with $T^\circ_{\tilde\rho,\tilde\psi}$ mapping $g \in L^\circ_p(A)$ to

$$
\begin{aligned}
(T^\circ_{\tilde\rho,\tilde\psi}g)(\sigma) &= \lambda_k^{-1/r} \rho(\theta_k^{-1}\sigma) \cdot \int_0^\sigma g(\tau)\,d\tau \\
&= \lambda_k^{-1/r} \rho(\theta_k^{-1}\sigma) \cdot (T_1 g)(\sigma), \quad \sigma \in A_k.
\end{aligned}
\tag{3.33}
$$

3.4. Transformations and Norms

It will be important that scale transformations preserve several norms of kernel functions appearing in our investigations. Suppose $1 < p \leq \infty$ and $1 \leq q \leq \infty$. Let ρ, ψ be non-negative functions on $(0, \infty)$ with $\psi \in L_{p'}(0, x)$ and $\rho \in L_q(x, \infty)$ for each $x > 0$. Let $u_k > 0$, $k \in \mathbb{Z}$, be the minimal real number which satisfies

$$
2^k = \int_0^{u_k} \psi(t)^{p'}\,dt.
\tag{3.34}
$$

If there is no solution of (3.34), we set $u_k := \infty$. Let

$$
\delta_k(\rho, \psi) := 2^{k/p'} \|\rho\|_{L_q(u_k, u_{k+1})} = \|\psi\|_{L_{p'}(u_k, u_{k+1})} \cdot \|\rho\|_{L_q(u_k, u_{k+1})}
\tag{3.35}
$$

and define the "norm"

$$
|(\rho, \psi)|_r := \left[\sum_{k \in \mathbb{Z}} \delta_k(\rho, \psi)^r \right]^{1/r}.
\tag{3.36}
$$

Note that we already used a special case of this construction, namely in (2.8), (2.9) we defined, in fact,

$$
|\rho|_r = |(\rho, \mathbf{1})|_r.
$$

Furthermore, we also need similar quantities to treat forward integration operators. If $\chi \in L_q(0, x)$ and $\eta \in L_{p'}(x, \infty)$ for each $x > 0$, let $v_k > 0$, $k \in$ be the maximal real number which satisfies

$$
2^k = \int_{v_k}^\infty \eta(t)^{p'}\,dt
$$

(and set $v_k := 0$ if there is no solution). Finally let

$$
\delta_k^*(\chi, \eta) := 2^{k/p'} \|\chi\|_{L_q(v_{k+1}, v_k)} = \|\eta\|_{L_{p'}(v_{k+1}, v_k)} \cdot \|\chi\|_{L_q(v_{k+1}, v_k)}
\tag{3.37}
$$

With this notation the following is valid.

PROPOSITION 3.11. *For $p > 1$ and $1 \leq q \leq \infty$ let ρ, ψ be non-negative functions with $\psi \in L_{p'}(0, x)$ and $\rho \in L_q(x, \infty)$ for each $x > 0$. Recall that the parameter r is defined by $1/r = 1/p' + 1/q > 0$.*

(1) *For any scale transformation θ from I onto A we have*

$$
\left\|\tilde\rho\tilde\psi\right\|_{L_r(A)} = \left\|\bar\rho\bar\psi\right\|_{L_r(I)} = \left\|\rho\psi\right\|_{L_r(I)}
$$

with $\tilde\rho$, $\tilde\psi$, $\bar\rho$ and $\bar\psi$ defined by (3.14), (3.15) and (3.13), respectively.

(2) *If a scale transformation θ on $(0, \infty)$ is increasing, then for $k \in \mathbb{Z}$ it follows that*

$$
\delta_k(\tilde\rho, \tilde\psi) = \delta_k(\bar\rho, \bar\psi).
$$

(3) *If a scale transformation θ on $(0, \infty)$ is decreasing, we have*

$$
\delta_k^*(\tilde\rho, \tilde\psi) = \delta_k(\bar\rho, \bar\psi).
$$

PROOF. (1) By the definition and by (3.8), it follows that

$$\left\|\tilde\rho\,\tilde\psi\right\|_r^r = \int_A |\rho(\theta^-\tau)\,\psi(\theta^-\tau)|^r\,\theta'(\theta^-\tau)^{-r/p'-r/q}\,d\tau$$

$$= \int_{N_\theta} |\rho(t)\,\psi(t)|^r\,dt = \left\|\bar\rho\,\bar\psi\right\|_r^r = \|\bar\rho\,\psi\|_r^r.$$

proving (1).

Let us now verify (2). Recall that by (3.9) for every $\omega \in A$

(3.38)
$$\int_0^\omega \tilde\psi(\tau)^{p'}d\tau = \int_0^{\theta^-\omega} \tilde\psi(\theta t)^{p'} \theta'(t)dt$$
$$= \int_0^{\theta^-\omega} \psi(\theta^-\theta t)^{p'}\theta'(\theta^-\theta t)^{-1}\theta'(t)dt = \int_0^{\theta^-\omega} \bar\psi(t)^{p'}dt,$$

where we used that θ' vanishes outside of N_θ and $\theta^-\theta t = t$ on N_θ. If ω_k is the minimal solution of the equation

$$2^k = \int_0^{\omega_k} \tilde\psi(\tau)^{p'}d\tau,$$

then by (3.38) the number $u_k := \theta^-\omega_k$ solves the similar equation

$$2^k = \int_0^{u_k} \bar\psi(t)^{p'}dt.$$

Moreover, u_k is its minimal solution. Indeed, if $\omega < \omega_k$, we have

$$2^k > \int_0^\omega \tilde\psi(\tau)^{p'}d\tau = \int_0^{\theta^-\omega} \bar\psi(t)^{p'}dt.$$

Since θ^- is left continuous, we obtain that for every $u < u_k$ there exists $\omega < \omega_k$ such that $u \le \theta^-\omega < u_k$, hence

$$\int_0^u \bar\psi(t)^{p'}dt \le \int_0^{\theta^-\omega} \bar\psi(t)^{p'}dt < 2^k.$$

Now using once more (3.9), by the definition of the u_k's and of $\tilde\rho$ we have

$$\int_{u_k}^{u_{k+1}} \bar\rho(t)^q\,dt = \int_{\omega_k}^{\omega_{k+1}} \tilde\rho(\tau)^q\,d\tau$$

which completes the proof of (2) for $q < \infty$. The case $q = \infty$ may be treated similarly.

Since (3) follows exactly in the same way, we omit the proof. □

Applications:
(1) $\theta(s) = \int_0^s \psi(t)^{p'}dt$ as in Example 3.3(a), see (3.22). Here we have $\tilde\psi \equiv 1$,

(3.39) $$\|\tilde\rho\|_r = \|\rho\,\psi\,\mathbf{1}_{\{\psi>0\}}\|_r = \|\rho\,\psi\|_r$$

and

(3.40) $$\delta_k(\tilde\rho,1) = \delta_k(\bar\rho,\psi) = \delta_k(\rho\,\mathbf{1}_{\{\psi>0\}},\psi).$$

Therefore

(3.41) $$|\tilde\rho|_r = |(\bar\rho,\psi)|_r.$$

(2) $\theta(s) = \int_s^\infty \rho(t)^q\, dt$ as in Example 3.3(c), see (3.28). Here we have $\tilde{\rho} \equiv 1$,
$$\|\tilde{\psi}\|_r = \|\rho\,\psi\|_r$$
and
$$\delta_k^*(\mathbf{1}, \tilde{\psi}) = \delta_k(\rho, \tilde{\psi}) = \delta_k(\rho, \psi\, \mathbf{1}_{\{\rho>0\}})\,.$$

(3) $\theta(s) = 1/s$ as in Example 3.3(b), see (3.25). Recall that here $\rho = \bar{\rho}$ as well as $\psi = \bar{\psi}$, and we have
$$\delta_k(\rho, \psi) = \delta_k^*(\tilde{\rho}, \tilde{\psi})$$
with $\tilde{\rho}, \tilde{\psi}$ defined in (3.26) and (3.27), respectively. In particular, if $\psi \equiv 1$, then
$$\delta_k(\rho) = \delta_k(\rho, \mathbf{1}) = \delta_k^*(\tilde{\rho}, \tau^{-2/p'}) = 2^{k/p'} \cdot ||\tilde{\rho}||_{L_q(2^{-k-1}, 2^{-k})}\,.$$

Remark: A question of interest is how $|(\rho, \psi)|_r$ changes by going to $\left|(\tilde{\psi}, \tilde{\rho})\right|_r$ for $\tilde{\rho}, \tilde{\psi}$ defined by (3.26) and (3.27) and by replacing $p \to q'$ and $q \to p'$, i.e. by going from the operator $T_{\rho,\psi}$ to its dual $T'_{\rho,\psi}$ (cf. Proposition 3.8). One may directly verify that each of these norms can be estimated against the other with suitable constants depending only on p, q.

CHAPTER 4

Upper Estimates for Entropy Numbers

Here we prove Theorem 2.2 and derive further upper bounds for other types of Volterra operators, namely for $T_{\rho,\psi}$ and $S_{\chi,\eta}$. The proof of Theorem 2.2 will be done in several steps. The first step provides us with a general bound which will be applied in different situations later on.

4.1. A General Bound Based on Partitions

In this section we show how each partition of the interval of integration yields an upper bound for the entropy numbers of the integral operator. As in example (e) of section 3.3, let $I \subseteq (0, \infty)$ be some interval with $I = \bigcup_{k \in \mathcal{K}} I_k$ for some disjoint bounded intervals I_k. Then $L_p^\circ(I)$ is defined by (3.31) as before. For $\rho : I \to [0, \infty)$ we now construct real numbers β_k, $k \in \mathcal{K}$, as follows.

$$(4.1) \quad \beta_k := \sup\left\{ \left(\int_{I_k} |\rho(t)(T_1 h)(t)|^q \, dt \right)^{1/q} : h \in L_p(I_k), \int_{I_k} h(t) \, dt = 0, \sup_{t \in I_k} |(T_1 h)(t)| \leq 1 \right\}.$$

In different words, β_k is the norm of the multiplication operator defined by ρ regarded on the image $T_1(L_p^\circ(I_k))$ endowed with the L_∞-norm and mapping into $L_q(I_k)$. Here $L_p^\circ(I_k)$ consists of functions h in $L_p(I_k)$ with $\int_{I_k} h(t) \, dt = 0$.

With this notation the following holds.

PROPOSITION 4.1. *Assume $1 < p \leq \infty$, $1 \leq q \leq \infty$ and define $r < \infty$ as before by $1/r = 1/p' + 1/q$. If*

$$(4.2) \quad \Lambda := \sum_{k \in \mathcal{K}} \beta_k^r \, |I_k|^{r/p'} < \infty,$$

then for $n \in \mathbb{N}$ and T_ρ°, the restriction of T_ρ to $L_p^\circ(I)$, we have

$$(4.3) \quad e_n(T_\rho^\circ : L_p^\circ(I) \to L_q(I)) \leq \Lambda^{1/r} \cdot e_n(T_1 : L_p[0,1] \to L_\infty[0,1]) \leq c \cdot \Lambda^{1/r} \, n^{-1}.$$

PROOF. We may clearly assume $\rho \cdot \mathbf{1}_{I_k} \not\equiv 0$. Consequently, the numbers β_k are non–zero, hence the same holds for the λ_k's defined by

$$(4.4) \quad \lambda_k := \beta_k^r \cdot |I_k|^{r/p'-1} = \beta_k^r \cdot |I_k|^{-r/q}.$$

With these λ_k's we construct now a piecewise linear map $\theta : I \to (0, \infty)$ by (3.29). Hence, for $A_k := \theta(I_k)$ we get

$$|A_k| = \lambda_k \cdot |I_k| = \beta_k^r \cdot |I_k|^{r/p'}$$

which implies

$$A := \theta(I) = \bigcup_{k \in \mathcal{K}} A_k = (0, \Lambda)$$

by the definition of Λ. In particular, the isometry Φ_p^θ maps $L_p(0,\Lambda)$ onto $L_p(I)$. Fix $k \in \mathcal{K}$ for a moment and let us apply formula (3.32) for p and ∞, hence $r = p'$, and with $\rho = \mathbf{1}_{I_k}$. Then this implies

$$[T_1 : L_p^\circ(I_k) \to L_\infty(I_k)] = \lambda_k^{-1/p'} \cdot \Phi_\infty^\theta \circ [T_1 : L_p^\circ(A_k) \to L_\infty(A_k)] \circ (\Phi_p^\theta)^{-1}$$

with isometries Φ_∞^θ and Φ_p^θ. By the definition of the β_k's this leads to

(4.5) $$\|T_\rho^\circ h\|_q \leq \beta_k \cdot \lambda_k^{-1/p'} \|T_1(\Phi_p^\theta)^{-1} h\|_{L_\infty(0,\Lambda)}$$

for each $h \in L_p^\circ(I_k)$. Assume first $q < \infty$. For each $f \in L_p^\circ(I)$ we apply (4.5) with $h := f \cdot \mathbf{1}_{I_k}$ and by summing over k we obtain

(4.6)
$$\|T_\rho^\circ f\|_q^q \leq \sum_{k \in \mathcal{K}} \beta_k^q \cdot \lambda_k^{-q/p'} \cdot \|T_1(\Phi_p^\theta)^{-1} f\|_{L_\infty(0,\Lambda)}^q$$
$$= \Lambda \cdot \|T_1(\Phi_p^\theta)^{-1} f\|_{L_\infty(0,\Lambda)}^q$$

which is a consequence of

$$\beta_k^q \cdot \lambda_k^{-q/p'} = \beta_k^{q - rq/p'} \cdot |I_k|^{r/p'} = \beta_k^r \cdot |I_k|^{r/p'}$$

in view of (4.4). Hence the sum in (4.6) gives Λ by (4.2). Before we proceed further, let us mention a general fact about entropy numbers.

LEMMA 4.2. *Let $T : E \to F_1$ and $S : E \to F_2$ be compact linear operators between Banach spaces E, F_1 and E, F_2, respectively. Assume that $\|Tx\|_{F_1} \leq \|Sx\|_{F_2}$ for all $x \in E$. Then this implies $e_n(T) \leq e_n(S)$ for all $n \in \mathbb{N}$.*

PROOF. Indeed, if $Sx_1, \ldots, Sx_N$, $\|x_j\| \leq 1$, forms an ε–net of $\{Sx : \|x\| \leq 1\}$ in F_2, then by the relation between the norms the same is valid for $Tx_1, \ldots, Tx_N$ and $\{Tx : \|x\| \leq 1\}$ in F_1. From this observation, we easily get the desired estimate of entropy numbers of T by those of S. $\square$

We now apply Lemma 4.2 to T_ρ° and to $\Lambda^{1/q} \cdot T_1 \circ (\Phi_p^\theta)^{-1}$. Then (4.6) implies

$$e_n(T_\rho^\circ : L_p^\circ(I) \to L_q(I)) \leq \Lambda^{1/q} \cdot e_n(T_1 \circ (\Phi_p^\theta)^{-1} : L_p^\circ(I) \to L_\infty(0,\Lambda))$$
$$\leq \Lambda^{1/q} \cdot e_n(T_1 : L_p(0,\Lambda) \to L_\infty(0,\Lambda))$$
(4.7) $$= \Lambda^{1/q + 1/p'} \cdot e_n(T_1 : L_p(0,1) \to L_\infty(0,1))$$

proving the first part of (4.3). Since by Theorem 2.1 for $p > 1$ always

(4.8) $$e_n(T_1 : L_p(0,1) \to L_\infty(0,1)) \leq c\, n^{-1}$$

with some $c > 0$ only depending on p, the second estimate in (4.3) also holds and the proof is completed for $q < \infty$.

If $q = \infty$ we infer from (4.5), instead of (4.6), a similar bound

$$\|T_\rho^\circ f\|_\infty \leq \sup_{k \in \mathcal{K}} \beta_k \cdot \lambda_k^{-1/p'} \cdot \|T_1(\Phi_p^\theta)^{-1} f\|_{L_\infty(0,\Lambda)}$$
$$= \|T_1(\Phi_p^\theta)^{-1} f\|_{L_\infty(0,\Lambda)} .$$

As in (4.7), Lemma 4.2 yields

$$e_n(T_\rho^\circ : L_p^\circ(I) \to L_\infty(I)) \leq e_n(T_1 \circ (\Phi_p^\theta)^{-1} : L_p^\circ(I) \to L_\infty(0,\Lambda))$$
$$\leq e_n(T_1 : L_p(0,\Lambda) \to L_\infty(0,\Lambda))$$
$$= \Lambda^{1/p'} \cdot e_n(T_1 : L_p(0,1) \to L_\infty(0,1))$$

and we are done using (4.8) and $p' = r$. □

Remark: The statement and the proof of Proposition 4.1 remain valid if we replace the operator T_ρ° by $T_\rho^* : L_p(I) \to L_q(I)$ defined by

$$T_\rho^* f(s) := T_\rho(f \mathbf{1}_{I_k})(s), \qquad s \in I_k,$$

and dropping the restriction $\int_{I_k} h(t)\, dt = 0$ in the definition of β_k, see (4.1).

4.2. Proof of Theorem 2.2 (1)

Recall that we have to verify

(4.9) $$\sup_n n\, e_n(T_\rho) \leq c \cdot |\rho|_r$$

for some $c = c(p) > 0$. This will done in several steps.
In a first step we treat the case $p = 1$ and $q < \infty$. Here we have $r = q$, and hence also

$$|\rho|_r = \|\rho\|_r = \|\rho\|_q .$$

Assuming temporarily $\rho > 0$ a.e., we apply Proposition 3.10 and obtain

$$\sup_n n\, e_n(T_\rho) = \sup_n n\, e_n(T_1)$$

with $T_1 : L_1(0, \|\rho\|_q^q) \to L_q(0, \|\rho\|_q^q)$. By a scaling argument,

$$e_n(T_1) = \|\rho\|_q \cdot e_n(T_1 : L_1(0,1) \to L_q(0,1)) .$$

Since by Theorem 2.1

$$e_n(T_1 : L_1(0,1) \to L_q(0,1)) \leq c\, n^{-1},$$

this completes the proof under the additional assumption $\rho > 0$.

For $\rho \geq 0$ let us choose $\rho_i \in L_q(0,\infty)$ such that $\rho_i \searrow \rho$ and $\rho_i > 0$ a.e.. Of course, $\rho_i \geq \rho$ gives

$$n\, e_n(T_\rho) \leq n\, e_n(T_{\rho_i}) ,$$

which implies by (4.9), being proved already for $\rho_i > 0$,

(4.10) $$\sup_n n\, e_n(T_\rho) \leq c \cdot \|\rho_i\|_q .$$

Taking in (4.10) the limit $i \to \infty$, estimate (4.9) holds for ρ as well. Hence we are done with $p = 1$.

In the second step we suppose $p > 1$ and $1 \leq q \leq \infty$. With $I = (0, \infty)$, $\mathcal{K} = \mathbb{Z}$ and $I_k = \Delta_k = [2^k, 2^{k+1})$ for $k \in \mathbb{Z}$ we define $L_p^\circ(0, \infty) = L_p^\circ$ as in (3.31), i.e. $f \in L_p^\circ$ whenever $f \in L_p$ and $\int_{\Delta_k} f(t)\, dt = 0$ for all $k \in \mathbb{Z}$. Let $L_p^c(0,\infty) = L_p^c$ be the complementary space given by

(4.11) $$L_p^c(0,\infty) := \left\{ f \in L_p(0,\infty) : f = \sum_{k \in \mathbb{Z}} \alpha_k \mathbf{1}_{\Delta_k} \text{ for some real } \alpha_k\text{'s} \right\}$$

and denote by $P^\circ : L_p \to L_p^\circ$ and $P^c : L_p \to L_p^c$ the corresponding canonical projections. As before, T_ρ° is defined as the restriction of T_ρ to L_p° and, similarly, T_ρ^c denotes the restriction of T_ρ to L_p^c. Then we have the representation

$$T_\rho = T_\rho^\circ P^\circ + T_\rho^c P^c ,$$

hence using the additivity of entropy numbers this implies

(4.12) $$e_{2n-1}(T_\rho) \leq \|P^\circ\|\, e_n(T_\rho^\circ) + \|P^c\|\, e_n(T_\rho^c) .$$

Thus, in order to verify (4.9), it suffices to estimate $\sup_n n\, e_n(T_\rho^\circ)$ and $\sup_n n\, e_n(T_\rho^c)$ by a multiple of $|\rho|_r$. Let us state and prove a slightly stronger result.

PROPOSITION 4.3. *If $p > 1$, $q \geq 1$ and $r > 0$ as above, then it follows that*

(4.13) $$\sup_n n\, e_n(T_\rho^\circ) \leq c \cdot |\rho|_r \quad \text{and}$$

(4.14) $$\sup_n n\, e_n(T_\rho^c) \leq c \cdot |\rho|_{r,\infty} .$$

Remark: Since $|\rho|_{r,\infty} \leq c\, |\rho|_r$, $\|P^c\| \leq 1$ and $\|P^\circ\| \leq 2$ assertions (4.13) and (4.14) combined with (4.12) complete the proof of (4.9).

PROOF OF (4.13): We use the general partition–based bound presented in section 4.1. For $I_k = \Delta_k$ we estimate the numbers β_k of (4.1) for $k \in \mathbb{Z}$ by

$$\beta_k \leq \|\rho\|_{L_q(\Delta_k)} .$$

Since $|\Delta_k| = 2^k$, this implies

$$\sum_{k \in \mathbb{Z}} \beta_k^r \cdot |\Delta_k|^{r/p'} \leq \sum_{k \in \mathbb{Z}} 2^{kr/p'} \cdot \|\rho\|_{L_q(\Delta_k)}^r = |\rho|_r^r ,$$

so by Proposition 4.1 we finally get

$$\sup_n n\, e_n(T_\rho^\circ) \leq c \cdot |\rho|_r$$

as asserted. $\square$

PROOF OF (4.14): First we introduce an isomorphism φ_p of $l_p(\mathbb{Z}) = l_p$ onto $L_p^c(0, \infty)$ by

(4.15) $$\varphi_p : (x_k)_{k \in \mathbb{Z}} \longrightarrow \sum_{k \in \mathbb{Z}} x_k |\Delta_k|^{-1/p} \mathbf{1}_{\Delta_k} .$$

It is easy to see that the operator $T_\rho^c \circ \varphi_p$ can be represented as

(4.16) $$T_\rho^c \circ \varphi_p = R_1 + R_2$$

where R_1 and R_2 map l_p into $L_q(0, \infty)$ as follows: For $x = (x_k)_{k \in \mathbb{Z}}$ let

(4.17) $$(R_1 x)(s) := \rho(s) \cdot x_k \cdot |\Delta_k|^{-1/p} (s - 2^k), \quad s \in \Delta_k, \quad \text{and}$$

(4.18) $$(R_2 x)(s) := \rho(s) \cdot \left(\sum_{j < k} x_j |\Delta_j|^{1/p'} \right), \quad s \in \Delta_k .$$

We start with the investigation of R_1 and define $\mathcal{D} : l_p \to l_q$ as diagonal operator mapping $(x_k)_{k \in \mathbb{Z}}$ to $(\delta_k\, x_k)_{k \in \mathbb{Z}}$ with δ_k's defined in (2.8). Furthermore, setting

$$g_k(s) := \|\rho\|_{L_q(\Delta_k)}^{-1} \cdot |\Delta_k|^{-1} (s - 2^k) \rho(s) \mathbf{1}_{\Delta_k}, \quad s > 0,$$

we have $\|g_k\|_q \leq 1$ and, moreover, the g_k's are disjointly supported. Consequently, the operator $Y : l_q(\mathbb{Z}) \to L_q(0, \infty)$ defined by

(4.19) $$Y(x) := \sum_{k \in \mathbb{Z}} x_k\, g_k, \quad x = (x_k)_{k \in \mathbb{Z}} \in l_q,$$

is bounded with $\|Y\| \leq 1$. Given $x \in l_q$ and $s > 0$, then

$$\begin{aligned}(Y\mathcal{D}x)(s) &= \sum_{k\in\mathbb{Z}} x_k |\Delta_k|^{1/p'} \|\rho\|_{L_q(\Delta_k)} \cdot g_k(s) \\ &= \sum_{k\in\mathbb{Z}} x_k |\Delta_k|^{1/p} (s - 2^k) \rho(s) \mathbf{1}_{\Delta_k}(s) \\ &= (R_1 x)(s),\end{aligned}$$

i.e. we have $R_1 = Y \circ \mathcal{D}$. From this we finally derive

(4.20) $$e_n(R_1) \leq \|Y\| \, e_n(\mathcal{D}) \leq e_n(\mathcal{D}).$$

In order to handle R_2, we introduce the auxiliary summation operator $V : l_p \to l_p$ defined by

(4.21) $$Vx := \left(\sum_{j<k} x_j \frac{|\Delta_j|^{1/p'}}{|\Delta_k|^{1/p'}} \right)_{k\in\mathbb{Z}}.$$

We claim that V is bounded. To verify this, take $x = (x_k)_{k\in\mathbb{Z}}$ in l_p and note that

(4.22) $$Vx = \sum_{l=1}^{\infty} 2^{-l/p'} s_l(x),$$

where $s_l(x) := (x_{k-l})_{k\in\mathbb{Z}}$ in $l_p(\mathbb{Z})$ denotes the element x shifted by l steps. Of course, $\|s_l(x)\|_p = \|x\|_p$, hence by (4.22) and the triangle inequality we arrive at

$$\|Vx\|_p \leq c \cdot \|x\|_p$$

with $c = \sum_{l=1}^{\infty} 2^{-l/p'}$, and V is bounded as claimed above.

Let us return to R_2 now. We define an isometric embedding j_q from l_q into $L_q(0, \infty)$ by setting

(4.23) $$j_q(e_k) = \frac{\rho \cdot \mathbf{1}_{\Delta_k}}{\|\rho \cdot \mathbf{1}_{\Delta_k}\|_q}$$

where e_k denotes the k-th unit vector in l_q. It is easy to verify that

(4.24) $$R_2 = j_q \circ \mathcal{D} \circ V$$

with diagonal operator $\mathcal{D} : l_p \to l_q$ defined as above. Thus (4.24) yields

(4.25) $$e_n(R_2) \leq \|V\| \, e_n(\mathcal{D}),$$

and it remains to estimate $e_n(\mathcal{D} : l_p \to l_q)$. To do so, we refer to Proposition 2 in [8], which asserts the following. Let p, q, r, v be some positive indices and $S : l_p \to l_q$ be the diagonal operator generated by a sequence (σ_i). Assume that $(\sigma_i) \in l_{r,v}$ and define an index u by the equation

$$\frac{1}{u} = \frac{1}{r} - \frac{1}{q} + \frac{1}{p}.$$

Suppose

(4.26) $$\frac{1}{r} > \left[\frac{1}{q} - \frac{1}{p} \right]_+,$$

then we have by [8] that $e_n(S) \in l_{u,v}$. Moreover, an inspection of the proof in [8] shows

$$\|(e_n(S))\|_{u,v} \leq c \, \|(\sigma_i)\|_{r,v} \tag{4.27}$$

with constant c depending on the indices involved.

In our situation, let $v = \infty$, and, as before, define r from (2.2). Then we have $u = 1$ and (4.26) reduces to the inequality $1 > [1/q - 1/p]_-$ which is true since $1/p \leq 1$ and we excluded the case $p = 1$ and $q = \infty$. Therefore (4.26) is verified. Now (4.27) yields

$$\sup_n n \, e_n(S) = \|(e_n(S))\|_{1,\infty} \leq c \, \|(\sigma_i)\|_{r,\infty}.$$

We apply this result to the diagonal operators $\mathcal{D}^+$ and $\mathcal{D}^-$ generated by $(\delta_k)_{k \geq 0}$ and $(\delta_k)_{k<0}$ separately and obtain

$$\sup_n n \, e_n(\mathcal{D} : l_p \to l_q) \leq c \cdot (\|(\delta_k)_{k<0}\|_{r,\infty} + \|(\delta_k)_{k\geq 0}\|_{r,\infty}) = c \cdot |\rho|_{r,\infty} \tag{4.28}$$

Combining (4.16), (4.20), (4.25) and (4.28) we finally obtain (4.14) via the additivity of entropy numbers. Now Proposition 4.3, and, consequently, also Theorem 2.2 (1) are proved. □

4.3. Proof of Parts (2) and (3) in Theorem 2.2

We begin with the proof of (2) in Theorem 2.2, i.e. we have to show

$$\limsup_{n \to \infty} n \, e_n(T_\rho) \leq c \cdot \|\rho\|_r$$

for $\rho \in L_q(I)$ and $I \subseteq (0, \infty)$ bounded. Let us say that φ on I is an interval step function (i.s.f.) whenever

$$\varphi = \sum_{k=1}^{N} \alpha_k \cdot \mathbf{1}_{I_k}$$

for some disjoint intervals $I_k \subseteq I$ and some α_k's in $\mathbb{R}$.

The next two lemmas show that arbitrary ρ's may be well approximated by i.s.f.'s.

LEMMA 4.4. *Assume* $0 < v \leq q < \infty$. *For each* $\rho \geq 0$ *in* $L_q(I)$, *I a bounded interval, and each* $\varepsilon > 0$ *we find an i.s.f.* $\varphi \geq 0$ *with* $\|\rho\|_v = \|\varphi\|_v$ *and such that*

$$\int_I [\rho(s)^q - \varphi(s)^q]_+ \, ds < \varepsilon.$$

The next lemma treats the case $q = \infty$. Recall that ρ^* was defined in (2.11).

LEMMA 4.5. *Let* $0 < v < \infty$ *and let* $\rho \geq 0$ *be a.e. bounded on the finite interval I. Then for each* $\varepsilon > 0$ *we find an i.s.f.* φ *such that* $\rho \leq \varphi$ *a.e., but*

$$\int_I \varphi(s)^v \, ds \leq \int_I \rho^*(s)^v \, ds + \varepsilon.$$

PROOFS OF LEMMAS 4.4 AND 4.5: Let us choose any increasing sequence of partitions of I, say $\mathcal{I}^l = \{I_1^l, \ldots, I_{k_l}^l\}$ with

$$\lim_{l \to \infty} \sup_{k \leq k_l} |I_k^l| = 0.$$

Then we define i.s.f.'s φ_l by

$$\varphi_l := \sum_{k \leq k_l} \alpha_k^l \mathbf{1}_{I_k^l}$$

where

$$\alpha_k^l := \left\{ |I_k^l|^{-1} \cdot \int_{I_k^l} \rho(s)^v \, ds \right\}^{1/v}, \quad q < \infty,$$

$$\alpha_k^l := \underset{s \in I_k^l}{\operatorname{ess\,sup}} \ \rho(s), \quad q = \infty.$$

Suppose first $q < \infty$. Since $\rho^v \in L_{q/v}(I) \subseteq L_1(I)$, the martingale convergence theorem implies $\varphi_l^v \to \rho^v$ a.e. on I, hence this also holds for the q-th powers and Lebesgue's Dominated Convergence Theorem yields

$$\lim_{l \to \infty} \int_I [\rho(s)^q - \varphi_l(s)^q]_+ \, ds = 0$$

as asserted.

Assume now $q = \infty$. Since the partitions are increasing, it follows that $\varphi_1 \geq \varphi_2 \geq \cdots$ and in view of the definition of ρ^* we also have $\varphi_l \geq \rho^* \geq \rho$ a.e. as asserted. On the other hand, for each $s \in I$ and $\delta > 0$ it follows

$$\varphi_l(s) \leq \underset{\{x:|s-x|<\delta\}}{\operatorname{ess\,sup}} \rho(x)$$

for l big enough, thus in view of the definition of ρ^* we get

$$\lim_{l \to \infty} \varphi_l(s) \leq \rho^*(s).$$

By the monotone convergence theorem

$$\lim_{l \to \infty} \int_I \varphi_l(s)^v \, ds = \int_I \rho^*(s)^v \, ds$$

which completes the proof of Lemma 4.5 by choosing l large enough. □

PROOF OF THEOREM 2.2 (2): If $p = 1$, then $|\rho|_r = \|\rho\|_r$, and by Theorem 2.2 (1) we even have

$$\sup_n n\, e_n(T_\rho) \leq c \cdot \|\rho\|_r$$

for arbitrary intervals I. So it remains to verify (2) for $p > 1$. Let us first treat the case $q < \infty$. Given $\varepsilon > 0$, according to Lemma 4.4 (for $v = r$) there exists an i.s.f. φ with $\|\varphi\|_r = \|\rho\|_r$ and

$$\int_I [\rho(s)^q - \varphi(s)^q]_+ \, ds < \varepsilon.$$

Let $I_1, \ldots, I_N$ be the partition of I generated by φ, i.e. φ is constant on each I_k. With respect to this partition we define now $L_p^\circ(I)$ and $L_p^c(I)$ as in (3.31) and (4.11) and as before P° and P^c denote the canonical projections from $L_p(I)$ onto these subspaces. Then $\operatorname{rank}(P^c) \leq N < \infty$, hence by the exponential decay of $e_n(P^c)$ as $n \to \infty$ (cf. (1.3.36) of [**10**]) we obtain

(4.29) $$\limsup_{n \to \infty} n\, e_n(T_\rho) \leq \|P^\circ\| \limsup_{n \to \infty} n\, e_n(T_\rho^\circ),$$

where as before T_ρ° is the restriction of T_ρ to $L_p^\circ(I)$. Hence it remains to estimate $e_n(T_\rho^\circ)$ suitably. Given $f \in L_p^\circ(I)$, we have

$$\begin{aligned}
\|T_\rho^\circ f\|_q^q &= \sum_{k=1}^N \int_{I_k} \rho(s)^q |(T_1 f)(s)|^q \, ds \\
&\leq \sum_{k=1}^N \int_{I_k} \varphi(s)^q |(T_1 f)(s)|^q \, ds + \int_I [\rho(s)^q - \varphi(s)^q]_+ \, ds \cdot \|T_1 f\|_{L_\infty(I)}^q \\
&\leq \|T_\varphi^\circ f\|_q^q + \varepsilon^q \cdot \|T_1 f\|_\infty^q ,
\end{aligned}$$

and hence

$$\begin{aligned}
\|T_\rho^\circ f\|_q &\leq \left(\|T_\varphi^\circ f\|_q^q + \varepsilon^q \cdot \|T_1 f\|_\infty^q \right)^{1/q} \\
(4.30) \qquad &\leq \|T_\varphi^\circ f\|_q + \varepsilon \cdot \|T_1 f\|_\infty = \|(T_\varphi^\circ, \varepsilon T_1) f\|_{L_q \oplus L_\infty}
\end{aligned}$$

where $(T_\varphi^\circ, \varepsilon T_1)$ maps $f \in L_p^\circ(I)$ to $(T_\varphi^\circ f, \varepsilon T_1 f) \in L_q(I) \oplus_1 L_\infty(I)$. It is not difficult to see that

$$e_{n+m-1}((T_\varphi^\circ, \varepsilon T_1)) \leq e_n(T_\varphi^\circ) + \varepsilon \cdot e_m(T_1) ,$$

thus using (4.30), by Lemma 4.2 we finally get for all $m, n \in \mathbb{N}$

$$\begin{aligned}
e_{n+m-1}(T_\rho^\circ : L_p^\circ(I) \to L_q(I)) &\leq e_n(T_\varphi^\circ : L_p^\circ(I) \to L_q(I)) \\
&\quad + \varepsilon \cdot e_m(T_1 : L_p(I) \to L_\infty(I)) .
\end{aligned}$$

Theorem 2.1 implies $e_m(T_1 : L_p(I) \to L_q(I)) \leq c(I) m^{-1}$, so by the last estimate we easily obtain

$$(4.31) \qquad \limsup_{n \to \infty} n \, e_n(T_\rho^\circ) \leq \limsup_{n \to \infty} n \, e_n(T_\varphi^\circ) + c(I) \cdot \varepsilon ,$$

and to complete the proof it remains to estimate $e_n(T_\varphi^\circ)$. For each $k = 1, \ldots, N$ let $\alpha_k \geq 0$ be the value of φ on I_k. Then for $h \in L_p^\circ(I_k)$ we get

$$\int_{I_k} \varphi(s)^q |(T_1 h)(s)|^q \, ds \leq \alpha_k^q \cdot |I_k| \cdot \sup_{s \in I_k} |(T_1 h)(s)|^q ,$$

i.e. the β_k's defined in (4.1) satisfy

$$\beta_k \leq \alpha_k \cdot |I_k|^{1/q} , \quad k = 1, \ldots, N .$$

Hence for Λ given by (4.2) we obtain

$$\Lambda = \sum_{k=1}^N \beta_k^r \cdot |I_k|^{r/p'} \leq \sum_{k=1}^N \alpha_k^r \cdot |I_k|^{r/p' + r/q} = \|\varphi\|_r^r = \|\rho\|_r^r$$

by the choice of φ. So Proposition 4.1 applies and leads to

$$(4.32) \qquad e_n(T_\varphi^\circ) \leq c \, \|\rho\|_r \, n^{-1} .$$

Finally, we combine (4.29) and (4.31) with (4.32) and obtain

$$\limsup_{n \to \infty} n \, e_n(T_\rho) \leq c \, \|\rho\|_r + c(I) \varepsilon$$

for any $\varepsilon > 0$. Letting $\varepsilon \to 0$ completes the proof of Theorem 2.2 (2) in the case $q < \infty$.

For $q = \infty$ we apply Lemma 4.5 with $v = r$. Thus for $\varepsilon > 0$ we find an i.s.f. φ such that $\|\varphi\|_r \leq \|\rho^*\|_r + \varepsilon$ and $\rho \leq \varphi$ a.e.. Consequently, an application of Proposition 4.1 to φ leads as before to

$$\limsup_{n \to \infty} n\, e_n(T_\rho^\circ) \leq \limsup_{n \to \infty} n\, e_n(T_\varphi^\circ) \leq c \cdot \|\varphi\|_r \leq c \cdot (\|\rho^*\|_r + \varepsilon) \, .$$

Then (4.29) completes the proof of Theorem 2.2 (2) for $q = \infty$. □

PROOF OF (3) IN THEOREM 2.2: For $p = 1$ part (3) is again a direct consequence of part (1). Therefore, let us suppose $p > 1$. Since $|\rho|_r < \infty$, for any given $\varepsilon > 0$ we can choose an integer K such that

$$(4.33) \qquad \left[\sum_{|k| \geq K} 2^{kr/p'} \left(\int_{\Delta_k} \rho(s)^q\, ds \right)^{r/q} \right]^{1/r} \leq \varepsilon \, .$$

Using this number K we define now functions ρ_1 and ρ_2 by

$$\rho_1 := \rho \cdot \mathbf{1}_{[2^{-K}, 2^K]} \quad \text{and} \quad \rho_2 := \rho - \rho_1 \, .$$

From (4.33) we derive $|\rho_2|_r \leq \varepsilon$, hence by Theorem 2.2 (1) this implies

$$\sup_n n\, e_n(T_{\rho_2}) \leq c \cdot \varepsilon \, .$$

On the other hand, note that for $I = [2^{-K}, 2^K]$ we have $\rho_1 \in L_q(I)$, consequently Theorem 2.2 (2) gives

$$\limsup_{n \to \infty} n\, e_n(T_{\rho_1}) \leq c \cdot \|\rho_1\|_r \leq c \cdot \|\rho\|_r \, , \quad q < \infty, \quad \text{and}$$

$$\limsup_{n \to \infty} n\, e_n(T_{\rho_1}) \leq c \cdot \|\rho_1^*\|_r \leq c \cdot \|\rho^*\|_r \, , \quad q = \infty.$$

Since $T_\rho = T_{\rho_1} + T_{\rho_2}$, the additivity of the entropy numbers yields

$$\limsup_{n \to \infty} n\, e_n(T_\rho) \leq c\, (\|\rho\|_r + \varepsilon)$$

(with natural modification for $q = \infty$) and this completes the proof of part (3) in Theorem 2.2 by taking the limit $\varepsilon \to 0$. □

4.4. Entropy Estimates for $T_{\rho,\psi}$

Our next objective is to extend Theorem 2.2 to the case of Volterra operators $T_{\rho,\psi}$ defined in (1.1) with arbitrary kernel functions ψ. Recall that Theorem 2.2 concerns the special case $\psi \equiv 1$.

THEOREM 4.6. *Let $1 < p \leq \infty$ and $1 \leq q \leq \infty$. For all functions ρ and ψ on $(0, \infty)$ let $|(\rho, \psi)|_r$ be defined as in (3.36). Then the following are valid.*

(1) *For $T_{\rho,\psi} : L_p(0, \infty) \to L_q(0, \infty)$ we have*

$$\sup_n n\, e_n(T_{\rho,\psi}) \leq c \cdot |(\rho, \psi)|_r \, .$$

(2) *If $I \subseteq (0, \infty)$ is a bounded interval and $\rho \in L_q(I)$, $\psi \in L_{p'}(I)$, then for $q < \infty$ we have*

$$(4.34) \qquad \limsup_{n \to \infty} n\, e_n(T_{\rho,\psi}) \leq c \cdot \|\rho\psi\|_r$$

with $1/r = 1/p' + 1/q$. If $q = \infty$, then the right hand side of (4.34) has to be replaced by $\|\rho^* \psi\|_{p'}$.

(3) If $|(\rho,\psi)|_r < \infty$ and $T_{\rho,\psi} : L_p(0,\infty) \to L_q(0,\infty)$, then it also follows that

(4.35) $$\limsup_{n\to\infty} n\, e_n(T_{\rho,\psi}) \le c \cdot \|\rho\psi\|_r$$

with $\|\rho^*\psi\|_{p'}$ on the right hand side of (4.35) for $q = \infty$.

PROOF. For $\psi > 0$ a.e., assertion (1) follows directly from Proposition 3.6, Proposition 3.11 and Theorem 2.2. Hence let $\psi \in L_{p'}(0,x)$, $x > 0$, with $\psi \ge 0$ and define as in Example 3.3(a) an increasing transformation θ by

(4.36) $$\theta(s) := \int_0^s \psi(t)^{p'}\, dt, \quad s > 0,$$

as well as numbers $u_k = \theta^-(2^k) > 0$, $k \in \mathbb{Z}$, i.e. the u_k's are the smallest numbers satisfying

(4.37) $$2^k = \int_0^{u_k} \psi(t)^{p'}\, dt.$$

As before, we set $u_k = \infty$ provided that (4.37) has no solution. Next we choose a function $\psi_0 > 0$ such that for all $k \in \mathbb{Z}$

(4.38) $$\int_0^{u_k} \psi_0(t)\, dt \le \min\{1, 2^{k-1}\}.$$

For $\varepsilon > 0$ let ψ_ε be defined by

$$\psi_\varepsilon(t)^{p'} := \psi(t)^{p'} + \varepsilon \cdot \psi_0(t), \quad t > 0,$$

and if $k \in \mathbb{Z}$, define $u_k(\varepsilon)$ by the equation

$$2^k = \int_0^{u_k(\varepsilon)} \psi_\varepsilon(t)^{p'}\, dt.$$

Note that $\psi_\varepsilon > 0$, hence if $u_k(\varepsilon)$ exists, it is uniquely determined. If our equation has no solution, we let $u_k(\varepsilon) = \infty$. Of course, the $u_k(\varepsilon)$'s increase as ε becomes smaller and, moreover, they tend to u_k as ε tends to zero. We claim now that

(4.39) $$u_{k-1} \le u_k(\varepsilon) \le u_k$$

for all $k \in \mathbb{Z}$ (eventually, with a single exception) and all ε with $0 < \varepsilon < 1$. If $u_k(\varepsilon) = \infty$, then so is u_k and there is nothing to prove. Hence let us assume $u_k(\varepsilon) < \infty$. We clearly have $u_k(\varepsilon) \le u_k$ by the construction of ψ_ε. On the other hand, since

$$\begin{aligned}2^k &= \int_0^{u_k(\varepsilon)} \psi(t)^{p'}\, dt + \varepsilon \int_0^{u_k(\varepsilon)} \psi_0(t)\, dt \\ &\le 2^k - \int_{u_k(\varepsilon)}^{u_k} \psi(t)^{p'}\, dt + \varepsilon \int_0^{u_k(\varepsilon)} \psi_0(t)\, dt,\end{aligned}$$

by (4.38) we get, if $u_k < \infty$,

$$\int_{u_k(\varepsilon)}^{u_k} \psi(t)^{p'}\, dt \le \varepsilon \int_0^{u_k(\varepsilon)} \psi_0(t)\, dt \le \varepsilon \int_0^{u_k} \psi_0(t)\, dt < 2^{k-1} = \int_{u_{k-1}}^{u_k} \psi(t)^{p'}\, dt$$

which implies $u_{k-1} \le u_k(\varepsilon)$ as claimed above. On the other hand, if $u_{k-1} = \infty$, then obviously

$$\int_0^\infty \psi_\varepsilon(t)^{p'}\, dt = \int_0^\infty \psi(t)^{p'}\, dt + \varepsilon \int_0^\infty \psi_0(t)\, dt < 2^{k-1} + 2^{k-1} = 2^k$$

i.e. $u_k(\varepsilon) = \infty$ and (4.39) also holds. The only case where (4.39) is not proved is the combination $u_{k-1} < \infty$ and $u_k = \infty$. But even in this situation (4.39) works at least with $k-1$ instead of k; anyway, for all k we have

$$u_{k-2} \leq u_k(\varepsilon) \leq u_k$$

and derive (observe also that $r/q < 1$)

$$\|\rho\|^r_{L_q(u_k(\varepsilon), u_{k+1}(\varepsilon))} \leq \|\rho\|^r_{L_q(u_{k-2}, u_{k-1})} + \|\rho\|^r_{L_q(u_{k-1}, u_k)} + \|\rho\|^r_{L_q(u_k, u_{k+1})},$$

thus Lebesgue's Dominated Convergence Theorem applies to the sequence

$$2^{kr/p'} \|\rho\|^r_{L_q(u_k(\varepsilon), u_{k+1}(\varepsilon))}$$

and implies

(4.40) $$\lim_{\varepsilon \to 0} |(\rho, \psi_\varepsilon)|_r = |(\rho, \psi)|_r .$$

Using $\psi_\varepsilon \geq \psi$ and $\psi_\varepsilon > 0$ it follows that

$$\sup_n n\, e_n(T_{\rho,\psi}) \leq \sup_n n\, e_n(T_{\rho,\psi_\varepsilon}) \leq c \cdot |(\rho, \psi_\varepsilon)|_r ,$$

and in view of (4.40), statement (1) follows now by letting $\varepsilon \to 0$.

Next we verify (2) for $q < \infty$. Assume in the first step $\psi > 0$ a.e.. Then we define θ by (4.36) and $\tilde{\rho}$ by (3.24). Since $\psi \in L_{p'}(I)$, the range A of θ is a finite interval and, moreover, by (3.8)

$$\int_A \tilde{\rho}(\tau)^q\, d\tau = \int_I \rho(t)^q\, dt < \infty .$$

Hence, Theorem 2.2 (2) applies to $\tilde{\rho}$ and A and leads to

(4.41) $$\limsup_{n \to \infty} n\, e_n(T_{\tilde{\rho}}) \leq c\, \|\tilde{\rho}\|_r .$$

Using Corollary 3.3 (recall that we assume $\psi > 0$ a.e.), from (3.39) and (4.41) we derive

$$\limsup_{n \to \infty} n\, e_n(T_{\rho,\psi}) = \limsup_{n \to \infty} n\, e_n(T_{\tilde{\rho}}) \leq c\, \|\tilde{\rho}\|_r = c\, \|\rho\,\psi\|_r$$

as asserted.

Now let $\psi \geq 0$. Then we choose $\psi_i \in L_{p'}(I)$, $\psi_i > 0$, tending monotonely from above to ψ. Since $\psi_i \geq \psi$, we have $e_n(T_{\rho,\psi}) \leq e_n(T_{\rho,\psi_i})$, and by the first step it follows that

$$\limsup_{n \to \infty} n\, e_n(T_{\rho,\psi}) \leq \limsup_{n \to \infty} n\, e_n(T_{\rho,\psi_i}) \leq c\, \|\rho\,\psi_i\|_r$$

for all i's. Letting $i \to \infty$ proves (2) for $q < \infty$ by the monotone convergence theorem.

For $q = \infty$ we have

$$\tilde{\rho}(\tau) = \rho(\theta^- \tau), \quad \tau \in A .$$

Hence, since θ^- is continuous for $\psi > 0$ a.e., by the definition of ρ^* it follows that

$$\widetilde{\rho^*}(\tau) = \rho^*(\theta^- \tau) = (\rho \circ \theta^-)^*(\tau) = (\tilde{\rho})^*(\tau) .$$

If we now apply Theorem 2.2 (2) for $q = \infty$, we finally get

$$\limsup_{n \to \infty} n\, e_n(T_{\rho,\psi}) = \limsup_{n \to \infty} n\, e_n(T_{\tilde{\rho}}) \leq c\, \|(\tilde{\rho})^*\|_{p'} = c\|\widetilde{\rho^*}\|_{p'} = c\, \|\rho^*\,\psi\|_{p'} .$$

This proves (2) for $q = \infty$ and $\psi > 0$ a.e.. The general case $\psi \geq 0$ follows as for $q < \infty$ by approximating ψ from above by $\psi_i > 0$.

Assertion (3) of Theorem 4.6 follows by similar arguments. The only, yet important, difference is that now we have to verify $|\tilde{\rho}|_r < \infty$ in order to apply (3) of Theorem 2.2. For $\psi > 0$ a.e. this follows from the assumption. Observe that (3.41) implies
$$|\tilde{\rho}|_r = |(\rho, \psi)|_r < \infty.$$
For $q < \infty$ by Theorem 2.2 (3) we therefore have
$$(4.42) \qquad \limsup_{n \to \infty} n\, e_n(T_{\rho,\psi}) \leq c\, \|\tilde{\rho}\|_r = \|\rho\,\psi\|_r.$$
If $q = \infty$, by the same arguments as above (4.42) holds with $\|\rho^* \psi\|_{p'}$ on the right hand side.

To treat the general case $\psi \geq 0$, it suffices to find a sequence $\psi_i > 0$ tending to ψ monotonely from above such that $|(\rho, \psi_i)|_r < \infty$. Then the proof can be finished (for all q) as before by the monotone convergence theorem. But such ψ_i's were constructed (there called ψ_ε's) in the proof of (1) in Theorem 4.6, and hence (3) is proved for all $\psi \geq 0$ and all q. □

COROLLARY 4.7. *If $\rho \cdot \psi = 0$ a.s. and $|(\rho, \psi)|_r < \infty$, then*
$$\lim_{n \to \infty} n\, e_n(T_{\rho,\psi}) = 0.$$

Remark: In particular, the preceding Corollary applies to weighted summation operators from $l_p(\mathbb{Z})$ to $l_q(\mathbb{Z})$ defined by
$$\Sigma_{a,b}(x) := \left(\alpha_k \sum_{j<k} \beta_j\, x_j\right)_{k \in \mathbb{Z}}$$
for $x = (x_k)_{k \in \mathbb{Z}} \in l_p$ and some sequences of non–negative numbers $a = (\alpha_k)_{k \in \mathbb{Z}}$ and $b = (\beta_k)_{k \in \mathbb{Z}}$. Compactness properties of those summation operators will be the subject of a separate paper (cf. [**13**]).

Our next objective is to treat the case $p = 1$ which is not covered by Theorem 4.6. Indeed, it displays some specific features. We begin with $\psi \equiv 1$.

PROPOSITION 4.8. *Suppose $1 \leq q < \infty$ and $\rho \in L_q(0, \infty)$. Then for $T_\rho : L_1(0, \infty) \to L_q(0, \infty)$ the following holds:*
$$(4.43) \qquad c_1 \cdot \|\rho\|_q \leq \liminf_{n \to \infty} n\, e_n(T_\rho) \leq \limsup_{n \to \infty} n\, e_n(T_\rho) \leq \sup_n n\, e_n(T_\rho) \leq c_2 \cdot \|\rho\|_q.$$

PROOF. For $\rho > 0$ this is a direct consequence of Proposition 3.10 and of properties of the ordinary integration operator, see Theorem 2.1. For general ρ's, the upper bound follows from Theorem 2.2 (1) (recall that $|\rho|_r = \|\rho\|_r$ for $p = 1$) and the lower bound is a consequence of Theorem 2.4 (2). □

For $p = 1$ general operators $T_{\rho,\psi}$ do not admit a reduction to T_ρ as done in Theorem 4.6. Recall that the restriction $p > 1$ was essential in order to construct the scale transformation θ in (3.22). Therefore, it is not surprising that the behaviour of $e_n(T_{\rho,\psi})$ is more complicated than that of T_ρ as described in (4.43). In fact, the next result shows that in contrast to Proposition 4.8 the assumption $\|\rho\,\psi\|_q < \infty$ does not always imply $\limsup_{n \to \infty} n\, e_n(T_{\rho,\psi}) < \infty$.

THEOREM 4.9. *Let $1 < q < \infty$, $\rho \in L_q(x, \infty)$, $\psi \in L_\infty(0, x)$, $x > 0$. Regarding $T_{\rho,\psi}$ as operator from L_1 to L_q the following are valid.*

(1) We have

(4.44) $$\sup_n n\, e_n(T_{\rho,\psi}) \leq c \cdot \left(\sum_{k \in \mathbb{Z}} 2^k \, \|\psi\|_{L_\infty(v_{k+1},v_k)}^q \right)^{1/q}$$

where the v_k's are maximal numbers satisfying

(4.45) $$\int_{v_k}^\infty \rho(t)^q\, dt = 2^k, \quad k \in \mathbb{Z}.$$

(2) Whenever the right hand side of (4.44) is finite, then

(4.46) $$\limsup_{n \to \infty} n\, e_n(T_{\rho,\psi}) \leq c \cdot \|\rho\, \psi^*\|_q$$

with ψ^* defined in (2.11).

(3) For any $q \in (1,\infty)$ there exist $\rho \in L_q(x,\infty)$, $x > 0$, and $\psi \in L_\infty(0,\infty)$ such that $\|\rho\, \psi\|_q < \infty$, yet

$$\limsup_{n \to \infty} n\, e_n(T_{\rho,\psi} : L_1(0,\infty) \to L_q(0,\infty)) = \infty.$$

Since the proof is based on duality arguments, we postpone it to section 4.7, which follows some material required on forward integration operators.

4.5. Proof of Theorem 2.3

PROOF OF PART (1): We first treat the case $q < \infty$. Here we are given a sequence of non–negative real numbers $(b_k)_{k=1}^\infty$ satisfying

$$\sum_{k=1}^\infty b_k^r = \infty \quad \text{as well as} \quad \sum_{k=1}^\infty b_k^q < \infty,$$

and we have to construct a function $\rho \in L_r(0,\infty)$ such that

$$b_k = 2^{k/p'} \left(\int_{2^k}^{2^{k+1}} \rho(s)^q ds \right)^{1/q}$$

for each $k \geq 1$ and

(4.47) $$\limsup_{n \to \infty} n\, e_n(T_\rho) = \infty.$$

Of course, we may assume $0 < b_k \leq 1$ for all k's because, if $b_k = 0$ for some k, then we put $\rho \equiv 0$ on $\Delta_k = [2^k, 2^{k+1})$. Next we divide $\mathbb{N}$ into disjoint subsets K_m, $m = 1, 2, \ldots$, such that

$$\lim_{m \to \infty} \sum_{k \in K_m} b_k^r = \infty.$$

For each fixed $m \in \mathbb{N}$ we choose a number $\nu_m > 0$ with

$$\nu_m > \max_{k \in K_m} b_k^{-1}.$$

By the choice of ν_m we find a sequence $(n_k)_{k \geq 1}$ of integers with

(4.48) $$\nu_m^r \leq \frac{n_k}{b_k^r} \leq 2\nu_m^r, \quad k \in K_m, \; m = 1, 2, \ldots.$$

Hence let $N_m := \sum_{k \in K_m} n_k$. Then (4.48) implies

(4.49) $$\nu_m^r \sum_{k \in K_m} b_k^r \leq N_m \leq 2 \cdot \nu_m^r \sum_{k \in K_m} b_k^r$$

for all $m \in \mathbb{N}$. Next, we split each interval $\Delta_k = [2^k, 2^{k+1})$, $k \in K_m$, into $3 n_k$ disjoint intervals as follows:

$$\Delta_k = \bigcup_{j=1}^{n_k} \left(A_{k,j}^+ \cup I_{k,j} \cup A_{k,j}^- \right) ,$$

with

(4.50) $$|I_{k,j}| := l_k \leq \frac{2^k \cdot b_k^q}{2 \cdot n_k} ,$$

and

(4.51) $$\left| A_{k,j}^+ \right| = \left| A_{k,j}^- \right| := L_k \geq \frac{2^k}{4 n_k}, \quad j = 1, \ldots, n_k .$$

Recall that we assumed $b_k \leq 1$, hence those intervals do always exist. Furthermore, and this is the important point of the construction, we order the intervals in the way that

$$A_{k,1}^+ \prec I_{k,1} \prec A_{k,1}^- \prec A_{k,2}^+ \prec \cdots \prec I_{k,n_k} \prec A_{k,n_k}^-$$

where $\prec$ denotes the natural ordering of disjoint intervals on the real line. Finally, we define functions $\rho_{k,j}, \rho_k$ and ρ by

$$\rho_{k,j} := \frac{2^{-k/p'} \cdot b_k}{n_k^{1/q} \cdot l_k^{1/q}} \cdot \mathbf{1}_{I_{k,j}}, \quad j = 1, \ldots, n_k,$$

$$\rho_k := \sum_{j=1}^{n_k} \rho_{k,j}$$

and

$$\rho := \sum_{m=1}^{\infty} \sum_{k \in K_m} \rho_k .$$

Then we obtain

(4.52) $$\|\rho_{k,j}\|_q = \frac{2^{-k/p'} \cdot b_k}{n_k^{1/q}} \quad \text{and}$$

(4.53) $$\|\rho_k\|_q = 2^{-k/p'} \cdot b_k .$$

Now we verify that ρ possesses the properties stated in Theorem 2.3.
(a) We first prove that

(4.54) $$\int_0^{\infty} s^{q/p'} \rho(s)^q ds < \infty .$$

But this is a direct consequence of (4.53) via

$$\sum_{k \in K_m} \int_0^{\infty} s^{q/p'} \rho_k(s)^q ds \leq 2^{q/p'} \cdot \sum_{k \in K_m} 2^{kq/p'} \|\rho_k\|_q^q = 2^{q/p'} \cdot \sum_{k \in K_m} b_k^q$$

since

$$\sum_{m=1}^{\infty} \sum_{k \in K_m} b_k^q < \infty .$$

Given $f \in L_p(0, \infty)$, Hölder's inequality implies

$$\|T_\rho f\|_q^q = \int_0^\infty \rho(s)^q \left| \int_0^s f(t)\, dt \right|^q ds$$
$$\leq \int_0^\infty \rho(s)^q \cdot s^{q/p'} \cdot \|f\|_p^q\, ds\,,$$

and hence by (4.54) T_ρ is a bounded operator from $L_p(0,\infty)$ into $L_q(0,\infty)$ as claimed in (a).

(b) For each $m \in \mathbb{N}$ by (4.50) and (4.53) we have

$$\sum_{k \in K_m} \|\rho_k\|_r^r = \sum_{k \in K_m} \|\rho_k\|_q^r \cdot (n_k \cdot l_k)^{r/p'}$$
$$\leq \sum_{k \in K_m} 2^{-kr/p'} b_k^r (2^k b_k^q)^{r/p'} = \sum_{k \in K_m} b_k^q\,,$$

proving $\rho \in L_r(0,\infty)$. Since $\|\rho_k\|_q = 2^{-k/p'} b_k \leq b_k$ we also get $\rho \in L_q(0,\infty)$ by $\sum_k b_k^q < \infty$.

(c) This is a direct consequence of (4.53). Note that

$$b_k = 2^{k/p'} \cdot \|\rho_k\|_q = \delta_k(\rho)$$

for each $k \in K_m$.

(d) We now will prove $\limsup_{n \to \infty} n\, e_n(T_\rho) = \infty$. Hence define functions $h_{k,j} \in L_p(0,\infty)$ by

$$h_{k,j} := L_k^{-1/p} \left(\mathbf{1}_{A_{k,j}^+} - \mathbf{1}_{A_{k,j}^-} \right),\quad 1 \leq j \leq n_k,\ k \in K_m\,.$$

Our special ordering of the intervals yields

$$T_\rho(h_{k,j}) = L_k^{1/p'} \cdot \rho_{k,j}\,,\quad 1 \leq j \leq n_k\,.$$

Next we construct operators

$$X_m\ :\ l_p^{N_m} \to L_p\Big(\bigcup_{k \in K_m} \Delta_k \Big)\quad \text{and}$$
$$Y_m\ :\ \text{span}\,\{ \mathbf{1}_{I_{k,j}} : j = 1, \ldots, n_k,\ k \in K_m \} \longrightarrow l_q^{N_m}$$

by

$$X_m(e_{k,j}) := h_{k,j}\quad \text{and}$$
$$Y_m^{-1}(e_{k,j}) := \frac{\rho_{k,j}}{\|\rho_{k,j}\|_q},\ j = 1,\ldots,n_k,\ k \in K_m\,,$$

where the $e_{k,j}$'s are the j-th unit vectors in $l_p^{n_k}$ or $l_q^{n_k}$, respectively. Note that $\|X_m\| \leq 2$ while Y_m is an isometry from $\text{span}\,\{ \mathbf{1}_{I_{k,j}} : j = 1, \ldots, n_k,\ k \in K_m \}$ endowed with the L_q–norm onto $l_q^{N_m}$. By the construction

(4.55) $$Y_m \circ T_\rho \circ X_m = \mathcal{D}_m$$

with diagonal operator $\mathcal{D}_m$ possessing diagonal elements $(\delta_{k,j})_{j=1,k\in K_m}^{n_k}$ with (use (4.51), (4.52) and (4.48))

$$\begin{aligned}\delta_{k,j} &= L_k^{1/p'} \cdot \|\rho_{k,j}\|_q \geq 4^{-1/p'} \frac{2^{k/p'} \cdot 2^{-k/p'} \cdot b_k}{n_k^{1/p'+1/q}} \\ &= 4^{-1/p'} \frac{b_k}{n_k^{1/r}} \geq c \cdot \nu_m^{-1} .\end{aligned}$$

By well–known estimates for the entropy numbers of identity operators (cf. [**38**], 12.2.1) we obtain

(4.56) $$e_N(id : l_p^N \to l_q^N) \geq c \cdot N^{1/q-1/p},$$

and hence

$$\begin{aligned}e_{N_m}(\mathcal{D}_m) &\geq c \cdot \nu_m^{-1} e_{N_m}(id : l_p^{N_m} \to l_q^{N_m}) \\ &\geq c \cdot \nu_m^{-1} \cdot N_m^{1/q-1/p} .\end{aligned}$$

Thus for $m \in \mathbb{N}$ by (4.55) and (4.49)

$$\begin{aligned}N_m \cdot e_{N_m}(T_\rho) &\geq N_m \|X_m\|^{-1} \|Y_m\|^{-1} \cdot e_{N_m}(\mathcal{D}_m) \\ &\geq c \cdot \nu_m^{-1} \cdot N_m^{1/r} \\ &\geq c \cdot \Big(\sum_{k \in K_m} b_k^r \Big)^{1/r} .\end{aligned}$$

Consequently (d) follows from the special choice of the sets K_m.

For $q = \infty$ only minor modifications are necessary. Here we have $r = p'$, and the assumptions are $\sum b_k^r = \infty$ as well as $\sup_k b_k < \infty$. We have to construct a function $\rho \in L_r(0,\infty)$ satisfying (4.47) such that

$$b_k = 2^{k/p'} \sup_{s \in I_k} \rho(s)$$

for $k \geq 1$. We repeat the construction of intervals and their lengths as for $q < \infty$, yet now we set

$$\rho_{k,j} = 2^{-k/p'} b_k \mathbf{1}_{I_{k,j}} .$$

Then we obtain

$$\|\rho_{k,j}\|_\infty = \|\rho_k\|_\infty = 2^{-k/p'} b_k ,$$

and for

$$\rho = \sum_{m=1}^\infty \sum_{k \in K_m} \rho_k$$

it follows that

$$\|\rho\|_\infty \leq \sup_k b_k < \infty .$$

Since

$$\sup_{s>0} \rho(s) s^{1/p'} = \sup_k \sup_{s \in \Delta_k} \rho(s) s^{1/p'} \leq 2^{1/p'} 2^{k/p'} \|\rho_k\|_\infty \leq 2^{1/p'} b_k ,$$

the operator T_ρ is bounded from $L_p(0,\infty)$ into $L_\infty(0,\infty)$. Of course, by the construction

$$b_k = 2^{k/p'} \cdot \|\rho_k\|_\infty = \delta_k(\rho)$$

and the diagonal elements $(\delta_{k,j})_{j=1,\, k\in K_m}^{n_k}$ of the diagonal operator $\mathcal{D}_m$ satisfy

$$\delta_{k,j} = L_k^{1/p'} \cdot \|\rho_{k,j}\|_\infty \geq 4^{-1/p'} \frac{2^{k/p'} \cdot 2^{-k/p'} \cdot b_k}{n_k^{1/p'}} = c \cdot \frac{b_k}{n_k^{1/r}}\ .$$

The rest of the proof can be done exactly as for $q < \infty$. □

Remark: In particular, Theorem 2.3 (1) implies the existence of ρ's in L_r with $\sup_n n\, e_n(T_\rho) = \infty$. Note that this may also be derived from Theorem 2.4 (1). Indeed, if such ρ's would not exist, then by (2.14), $\|\rho\|_r < \infty$ would always imply $|\rho|_{r,\infty} < \infty$, which, as can be seen easily, is not valid.

PROOF OF THEOREM 2.3 (2): This follows along the same lines as the preceding proof. Fix $n \in \mathbb{N}$ and choose real numbers $l_n \leq 1/n$ (specified later on) and set $L_n = (1/n - l_n)/2$. We have

$$L_n \geq 1/4n \quad \text{and} \quad 2L_n + l_n = 1/n\ .$$

Next we divide $[0,1]$ into $3n$ disjoint intervals $(A_j^+)_{j=1}^n$, $(A_j^-)_{j=1}^n$ and $(I_j)_{j=1}^n$ as follows:

$$|A_j^+| = |A_j^-| = L_n \quad \text{and} \quad |I_j| = l_n\ , \quad j = 1,\ldots, n,$$

and, as above,

$$A_1^+ \preceq I_1 \preceq A_1^- \preceq \cdots \preceq A_n^+ \preceq I_n \preceq A_n^-\ .$$

Now we set

$$\rho_j := l_n^{-1/q} \cdot \mathbf{1}_{I_j} \quad \text{and} \quad \rho := \sum_{j=1}^n \rho_j$$

and obtain

$$\|\rho_j\|_q = 1\ , \quad \|\rho_j\|_r = l_n^{1/p'} \quad \text{and} \quad \|\rho\|_r = n^{1/r} \cdot l_n^{1/p'}\ .$$

Let

$$h_{n,j} := L_n^{-1/p}\left(\mathbf{1}_{A_j^+} - \mathbf{1}_{A_j^-}\right),\quad 1 \leq j \leq n.$$

By the construction,

$$T_\rho(h_{n,j}) = L_n^{1/p'} \cdot \rho_j\ ,$$

and using (4.56) with $N = n$, we obtain

$$e_n(T_\rho) \geq c\, L_n^{1/p'}\, \|\rho_j\|_q\, n^{1/q - 1/p} \geq c \cdot n^{1/q - 1}\ .$$

Therefore

(4.57) $$\|\rho\|_r^{-1} \cdot n\, e_n(T_\rho) \geq c\, (n\, l_n)^{-1/p'}.$$

For M mentioned in the statement of the theorem, we choose l_n so small that

$$c \cdot (n\, l_n)^{-1/p'} \geq M\ ,$$

and by (4.57) this leads to

$$n\, e_n(T_\rho) \geq M \cdot \|\rho\|_r$$

as asserted. □

4.6. Upper Bounds for Forward Integration Operators

Now we provide upper bounds for entropy numbers of operators $S_{\chi,\eta}$ defined in (3.21).

THEOREM 4.10. *Let $p > 1$ and let χ and η be as in (3.20). Assume $S_{\chi,\eta} : L_p(0,\infty) \to L_q(0,\infty)$ is a bounded forward integration operator.*

(1) *It follows that*

$$\sup_n n\, e_n(S_{\chi,\eta}) \leq c \left(\sum_{k \in \mathbb{Z}} \delta_k^*(\chi,\eta)^r \right)^{1/r} \tag{4.58}$$

with $\delta_k^*(\chi,\eta)$ defined in (3.37).

(2) *If we have $\chi, \psi \equiv 0$ on $(0,x)$ for some $x > 0$ and $\chi \in L_q(x,\infty)$ as well as $\eta \in L_{p'}(x,\infty)$, then for $q < \infty$*

$$\limsup_{n \to \infty} n\, e_n(S_{\chi,\eta}) \leq c \cdot \|\chi\, \eta\|_r \tag{4.59}$$

while for $q = \infty$ we have

$$\limsup_{n \to \infty} n\, e_n(S_{\chi,\eta}) \leq c \cdot \|\chi^*\, \eta\|_{p'} . \tag{4.60}$$

(3) *Whenever the right hand side of (4.58) is finite, then (4.59) (or (4.60) for $q = \infty$) also hold.*

PROOF. We know from Proposition 3.7 that $S_{\chi,\eta}$ is isomorphic to $T_{\tilde{\chi},\tilde{\eta}}$ with $\tilde{\chi},\tilde{\eta}$ defined therein. Consequently, for $q < \infty$ assertions (1), (2) and (3) follow by a straightforward application of Theorem 4.6 (1), (2) and (3) to $T_{\tilde{\chi},\tilde{\eta}}$ by using the equalities $\delta_k^*(\chi,\eta) = \delta_k(\tilde{\chi},\tilde{\eta})$ and $\|\chi\,\eta\|_r = \|\tilde{\chi}\,\tilde{\eta}\|_r$ as proved in Proposition 3.11. Recall that $S_{\chi,\eta}$ and $T_{\tilde{\chi},\tilde{\eta}}$ are isomorphic, thus they possess the same entropy behaviour.

If $q = \infty$, some additional consideration is necessary. Note that for the transformation θ defined by (3.25) the inverse $\theta^- = \theta^{-1}$ is continuous. Hence, by the same arguments as in the proof of Theorem 4.6 (2) this implies $(\tilde{\chi})^* = \widetilde{\chi^*}$, from which one easily derives (4.60) via Theorem 4.6 (2) and (3), respectively. □

4.7. Proof of Theorem 4.9

Our arguments rest upon the following special case of a general result due to Bourgain et al. (cf. [5]).

PROPOSITION 4.11. *Let E and F be Banach spaces such that at least one of them is uniformly convex. Let $T: E \to F$ be a compact operator with dual operator $T': F' \to E'$.*

(1) *There exist universal constants c_1, c_2 (depending only on the modulus of convexity of the involved uniformly convex space) such that*

$$c_1 \sup_n n\, e_n(T) \leq \sup_n n\, e_n(T') \leq c_2 \sup_n n\, e_n(T) .$$

(2) *Similarly, if $\sup_n n\, e_n(T) < \infty$ (or equivalently $\sup_n n\, e_n(T') < \infty$), then*

$$c_1 \limsup_{n \to \infty} n\, e_n(T) \leq \limsup_{n \to \infty} n\, e_n(T') \leq c_2 \limsup_{n \to \infty} n\, e_n(T) .$$

PROOF. Since (2) is not explicitly in [**5**], let us briefly indicate how it can be derived from the results stated there. If E is uniformly convex and $T: E \to F$, then in the proof of Corollary 2 in [**5**] the following estimates can be found.

$$(4.61) \qquad e_{2n}(T) \le c \cdot e_n(T)^{(\gamma-1)/\gamma} \cdot e_n(T')^{1/\gamma} \quad \text{and}$$
$$(4.62) \qquad e_{2n}(T') \le c \cdot e_n(T')^{\gamma/(\gamma+1)} \cdot e_n(T)^{1/(\gamma+1)} .$$

Here $\gamma \in [2, \infty)$ as well as $c > 0$ only depend on the modulus of convexity of E. For F uniformly convex the above estimates hold after replacing T by T' and vice versa. Of course, then γ and $c > 0$ depend on the modulus of convexity of F. From (4.61) we easily get

$$\limsup_{n\to\infty} (2n) e_{2n}(T) \le c \cdot \limsup_{n\to\infty} n^{(\gamma-1)/\gamma} e_n(T)^{(\gamma-1)/\gamma} \cdot \limsup_{n\to\infty} n^{1/\gamma} e_n(T')^{1/\gamma}$$
$$(4.63) \qquad = c \cdot \left(\limsup_{n\to\infty} n \, e_n(T) \right)^{(\gamma-1)/\gamma} \cdot \left(\limsup_{n\to\infty} n \, e_n(T') \right)^{1/\gamma} .$$

The monotonicity of $e_n(T)$ implies

$$(4.64) \qquad \limsup_{n\to\infty} n \, e_n(T) = \limsup_{n\to\infty} (2n) \, e_{2n}(T) ,$$

hence, since by assumption $\limsup_{n\to\infty} n \, e_n(T) < \infty$, estimates (4.63) and (4.64) yield

$$\limsup_{n\to\infty} n \, e_n(T) \le c \cdot \limsup_{n\to\infty} n \, e_n(T') .$$

The opposite inequality follows exactly by the same arguments starting with (4.62) instead of (4.61). □

PROOF OF THEOREM 4.9 : Since $T_{\rho,\psi} : L_1(0,\infty) \to L_q(0,\infty)$, its dual $S_{\psi,\rho}$ maps from $L_{q'}$ to L_∞, and because $L_q(0,\infty)$ is uniformly convex for $1 < q < \infty$, Proposition 4.11 applies. Thus $\sup_n n \, e_n(T_{\rho,\psi})$ and $\sup_n n \, e_n(S_{\psi,\rho})$ as well as their upper limits can be estimated one against the other with constants depending only on q. Because of $q' > 1$ we are in the situation of Theorem 4.10 and obtain from (4.58)

$$(4.65) \qquad \sup_n n \, e_n(S_{\psi,\rho}) \le c \cdot \left(\sum_{k \in \mathbb{Z}} \delta_k^*(\psi, \rho)^r \right)^{1/r}$$

where

$$\delta_k^*(\psi, \rho) = 2^{k/q} \cdot \|\psi\|_{L_\infty(v_{k+1}, v_k)}$$

with v_k's defined by (4.45). Thus by $q = r$, assertion (1) follows from (4.65).

Assertion (2) can be derived by similar arguments from (4.60) noting that $\|\rho \, \psi^*\|_r = \|\rho \, \psi^*\|_q$.

To verify assertion (3) we transform $S_{\psi,\rho}$ isomorphically "back" to the operator $T_{\tilde\psi,\tilde\rho} : L_{q'} \to L_\infty$ as in Example 3.3(b), Proposition 3.7, with

$$\tilde\psi(\tau) = \psi(\tau^{-1}), \quad \tau > 0, \quad \text{and}$$
$$\tilde\rho(\tau) = \rho(\tau^{-1}) \cdot \tau^{-2/q}, \quad \tau > 0 .$$

Next we choose special functions ρ and ψ. Defining ρ by

$$\rho(t) := t^{-2/q}, \quad t > 0,$$

we have $\rho \in L_q(x, \infty)$, $x > 0$, and, moreover, $\tilde{\rho} \equiv 1$. To choose ψ let ρ_0 be the function constructed in Theorem 2.3 (1) for certain b_k's now satisfying

$$\sum_{k=1}^{\infty} b_k^q = \infty \quad \text{and} \quad \sup_k b_k < \infty .$$

Then we have $\|\rho_0\|_q < \infty$ as well as $\|\rho_0\|_\infty < \infty$, yet for the operator T_{ρ_0} from $L_{q'}$ to L_∞ we have

$$\limsup_{n \to \infty} n \, e_n(T_{\rho_0}) = \infty .$$

With this function ρ_0 we put

$$\psi(t) := \rho_0(t^{-1}), \quad t > 0 .$$

Of course, $\psi \in L_\infty(0, \infty)$ and $\tilde{\psi} = \rho_0$, hence

$$T_{\tilde{\psi}, \tilde{\rho}} = T_{\rho_0}$$

and, by the construction,

$$\limsup_{n \to \infty} n \, e_n(T_{\tilde{\psi}, \tilde{\rho}}) = \infty .$$

Consequently the same holds for $S_{\psi, \rho}$ and in view of Proposition 4.11 also for $T_{\rho, \psi}$ as asserted. Thus it remains to prove $\|\rho \psi\|_q < \infty$. But this follows from

$$\int_0^\infty \rho(t)^q \, \psi(t)^q \, dt = \int_0^\infty t^{-2} \cdot \rho_0(t^{-1})^q \, dt = \|\rho_0\|_q^q < \infty$$

and completes the proof of assertion (3). □

Remark: The totality of our results covers all other admissible combinations of p and q except the case $p = q = 1$ (recall that we always exclude $p = 1$, $q = \infty$ leading to non–compact Volterra operators). We found sufficient conditions for $\sup_n n \, e_n(T_{\rho, \psi}) < \infty$ and they are best possible among integrability conditions. The case $p = q = 1$ remains open because here Proposition 4.11 does not apply, i.e. we are unable to transform entropy estimates for operators from L_∞ to L_∞ into those for operators in L_1. But also in this case some partial result is available. Namely, if we combine Theorem 6.23 from Chapter 6 below with estimate (2.1), then it follows, that at least assertion (1) of Theorem 4.9 is also valid for operators from $L_1(0, \infty)$ to $L_1(0, \infty)$.

CHAPTER 5

Lower Estimates for Entropy Numbers

The aim of this chapter is to prove Theorem 2.4 and to obtain further lower bounds for other classes of Volterra operators. As in the case of upper estimates, we start with a quite general construction from which we later derive various lower estimates.

5.1. A General Construction for Lower Estimates

To prove lower estimates for $e_n(T_\rho)$ we have to approximate ρ by interval step functions (i.s.f.'s) $\varphi \geq 0$ from below. Yet, in general, not every non–negative measurable function ρ on $(0,\infty)$ can be approximated in this way. It may even happen that there is no non–trivial i.s.f. φ with $0 \leq \varphi \leq \rho$. To overcome these difficulties, we have to weaken the measure of approximation. This will be made more precise in the next lemma which may be viewed as counterpart to Lemma 4.4 in the case of lower approximation.

LEMMA 5.1. *Let $I \subset (0,\infty)$ be a finite interval and let v be a positive number. Then for each $\rho \geq 0$ in $L_v(I)$ and all $\varepsilon, \delta > 0$ we find an i.s.f. φ such that the following holds:*
If

(5.1) $$G := \{s \in I : \rho(s) > (1-\delta)\varphi(s)\} ,$$

then

(5.2) $$\int_G \varphi(s)^v \, ds \geq \int_I \rho(s)^v \, ds - \varepsilon .$$

PROOF. Since $\rho^v \in L_1(I)$, as in the proof of Lemma 4.4 (with $v = q$) the martingale convergence theorem implies the existence of an i.s.f. $\varphi \geq 0$ such that $\|\rho\|_v = \|\varphi\|_v$ and, moreover,

(5.3) $$\int_I |\rho(s)^v - \varphi(s)^v| \, ds \leq \varepsilon \left(1 - (1-\delta)^v\right) .$$

For G given by (5.1) this implies

(5.4) $$\begin{aligned} \int_G \varphi(s)^v \, ds &= \int_I \varphi(s)^v \, ds - \int_{G^c} \varphi(s)^v \, ds \\ &= \int_I \rho(s)^v \, ds - \int_{G^c} \varphi(s)^v \, ds . \end{aligned}$$

On the other hand, if $s \notin G$, then

$$\rho(s)^v \leq (1-\delta)^v \cdot \varphi(s)^v .$$

Hence

$$[1 - (1-\delta)^v] \cdot \varphi(s)^v \leq \varphi(s)^v - \rho(s)^v \leq |\varphi(s)^v - \rho(s)^v| ,$$

and using (5.3), this implies
$$\int_{G^c} \varphi(s)^v \, ds \leq (1-(1-\delta)^v)^{-1} \cdot \int_I |\varphi(s)^v - \rho(s)^v| \, ds \leq \varepsilon \, .$$
Combining the last estimate with (5.4) finally leads to
$$\int_G \varphi(s)^v \, ds \geq \int_I \rho(s)^v \, ds - \varepsilon$$
as asserted. □

Our next objective is to transform an i.s.f. φ on I in such a way that for suitable partitions of I and $[0,1]$ the operator $T_\varphi : L_p^\circ(I) \to L_q(I)$ becomes isomorphic to a multiple of T_1° on $[0,1]$. More precisely, suppose

(5.5)
$$\varphi = \sum_{k=1}^N \alpha_k \cdot \mathbf{1}_{I_k}$$

for some disjoint I_k's in I and non–negative α_k's. Let $L_p^\circ(I)$ be defined by (3.31) with respect to the partition $I_1, \ldots, I_N$ and let the A_k's be disjoint intervals in $[0,1]$ with lengths

$$|A_k| = \|\varphi\|_r^{-r} \cdot \alpha_k^r \cdot |I_k| \, , \quad k=1,\ldots,N \, .$$

For $L_p^\circ(0,1)$ defined by these intervals the following is valid.

LEMMA 5.2. *The operator $T_\varphi^\circ : L_p^\circ(I) \to L_q(I)$ is isomorphic to the operator $\|\varphi\|_r \cdot T_1^\circ : L_p^\circ(0,1) \to L_q(0,1)$.*

PROOF. This is a direct consequence of formula (3.33) with
$$\lambda_k := \alpha_k^r \cdot \|\varphi\|_r^{-r} \, , \quad k = 1, \ldots, N \, .$$
□

For later purposes we have to generalize Lemma 5.2 slightly. So let $G \subseteq I$ be some measurable subset and for φ given by (5.5) we now define

(5.6)
$$\varphi_G := \varphi \cdot \mathbf{1}_G \, ,$$

which, of course, is no longer an i.s.f.. Yet the transformation presented in Example 3.3(d) allows us to assign to φ_G an i.s.f. $\tilde\varphi_G$ defined on a smaller interval. For $L_p^\circ(I)$ constructed via $I_1, \ldots, I_N$, the following holds.

PROPOSITION 5.3. *Let $1 \leq p,q \leq \infty$ be as before and let φ_G be defined by (5.6) with i.s.f. φ given by (5.5). Then*
$$T_{\varphi_G, \mathbf{1}_G}^\circ : L_p^\circ(I) \to L_q(I)$$
is isomorphic to
$$T_{\tilde\varphi_G}^\circ : L_p^\circ(0,|G|) \to L_q(0,|G|) \, ,$$
where $\tilde\varphi_G$ is an i.s.f. on $(0,|G|)$ of the form
$$\tilde\varphi_G = \sum_{k=1}^N \alpha_k \cdot \mathbf{1}_{G_k}$$
with $|G_k| = |I_k \cap G|$, $k = 1, \ldots, N$, and $L_p^\circ(0,|G|)$ defined with respect to the partition $G_1, \ldots, G_N$.

PROOF. This is a direct consequence of Example 3.3(d) applied with $\rho = \varphi$ and $\psi = \mathbf{1}_G$. □

If we combine the preceding results, we obtain the announced general construction for getting lower estimates of $e_n(T_\rho)$ and of other related quantities of T_ρ. Loosely speaking, we show that our operator T_ρ is "almost contained" in the operator $\|\rho\|_r \cdot T_1 : L_p(0,1) \to L_q(0,1)$, or, better said, a restriction of T_ρ to a finite-codimensional subspace of $L_p(I)$ is contained in a restriction of $\|\rho\|_r \cdot T_1$ to a finite-codimensional subspace of $L_p(0,1)$.

PROPOSITION 5.4. *Let $I \subset (0, \infty)$ be a finite interval and let $\rho \in L_q(I)$ be non-negative. Given $\varepsilon, \delta > 0$ there exist finite-codimensional subspaces $E \subseteq L_p(I)$ and $F \subseteq L_p(0,1)$, operators X and Y acting on $L_p(I)$ and $L_q(I)$, respectively, as well as a positive number λ possessing the following properties.*

(1) *We have*
$$\|X : L_p(I) \to L_p(I)\| \leq 1 \quad \text{and} \quad \|Y : L_q(I) \to L_q(I)\| \leq (1-\delta)^{-1}\;.$$

(2) *The number λ satisfies*
$$\lambda \geq (\|\rho\|^r - \varepsilon)^{1/r}\;.$$

(3) *The restriction of $Y \circ T_\rho \circ X$ to E is isomorphic to the restriction of $\lambda \cdot T_1$ to F with T_1 regarded as operator from $L_p(0,1)$ to $L_q(0,1)$.*

PROOF. For $\varepsilon, \delta > 0$ and $v = r$ let φ be an i.s.f. possessing the properties of Lemma 5.1. If $G \subset I$ is defined by (5.1), let φ_G be given by (5.6). It follows from (5.1) and (5.2) that
$$\rho \geq (1-\delta) \cdot \varphi_G \quad \text{and} \quad \|\varphi_G\|_r^r \geq \|\rho\|_r^r - \varepsilon\;.$$

Next we define the operators X and Y by
$$X(f) := \mathbf{1}_G \cdot f \quad \text{for} \quad f \in L_p(I)$$
and
$$(Yh)(s) := h(s) \cdot \frac{\varphi(s)}{\rho(s)} \cdot \mathbf{1}_G(s) = h(s) \cdot \frac{\varphi_G(s)}{\rho(s)} \quad \text{for} \quad h \in L_q(I) \text{ and } s \in I\;.$$

Note that $\rho(s) > 0$ for $s \in G$. Hence Y is well-defined and, of course,
$$\|X : L_p(I) \to L_p(I)\| \leq 1 \quad \text{and} \quad \|Y : L_q(I) \to L_q(I)\| \leq (1-\delta)^{-1}$$
by the special choice of the set G. Moreover, by the construction of X and Y it follows that
$$Y \circ T_\rho \circ X = T_{\varphi_G, \mathbf{1}_G}\;.$$

If $E := L_p^\circ(I)$ with respect to the partition generated by φ, then it is finite-codimensional in $L_p(I)$. Similarly, we define a finite-codimensional space $F \subseteq L_p(0,1)$ by $F := L_p^\circ(0,1)$ where this time we take the partition of $[0,1]$ generated by $A_1, \ldots, A_N$ with
$$|A_k| = \alpha_k^r \cdot |G \cap I_k| \cdot \|\varphi_G\|_r^{-r}\;, \quad k = 1, \ldots, N\;,$$
where as before the α_k's are the values of φ on the I_k's. Of course,
$$T_{\varphi_G, \mathbf{1}_G}^\circ = Y \circ T_\rho \circ X \Big|_E\;,$$

and therefore Proposition 5.3 applies and asserts that $Y \circ T_\rho \circ X\big|_E$ is isomorphic to $T_{\tilde{\varphi}_G}$. Moreover,
$$\|\varphi_G\|_r \cdot T_1^\circ = \|\varphi_G\|_r \cdot T_1\big|_F$$
with T_1 mapping $L_p(0,1)$ into $L_q(0,1)$, which by Lemma 5.2 is isomorphic to $T_{\tilde{\varphi}_G}$. Finally, we set $\lambda := \|\varphi_G\|_r$ and obtain
$$\lambda^r = \|\varphi_G\|_r^r \geq \|\rho\|_r^r - \varepsilon$$
which completes the proof. □

Remark: For later purposes (to obtain more precise constants) let us mention that there is a projection $P (= P^\circ)$ from $L_p(0,1)$ onto the finite-codimensional space F with $\|P\| \leq 2$. Moreover, if $p = 2$, then, of course, $\|P\| \leq 1$.

5.2. Proof of Theorem 2.4

We start with the proof of part (1). The basic idea is to construct suitable diagonal operators $\mathcal{D}^i$ with
$$e_n(\mathcal{D}^i) \leq c \cdot e_n(T_\rho),$$
and then apply lower estimates for entropy numbers of diagonal operators as proved in [8]. To be more precise, let $\delta_k = \delta_k(\rho)$ be defined by (2.8), i.e.
$$\delta_k = |\Delta_k|^{1/p'} \|\rho\|_{L_q(\Delta_k)}$$
for $k \in \mathbb{Z}$. These numbers generate three diagonal operators $\mathcal{D}^i : l_p \to l_q$, $i = 0, 1, 2$ as follows:
$$\mathcal{D}^i : (x_k)_{k \in \mathbb{Z}} \to (\delta_{3k+i} x_k)_{k \in \mathbb{Z}}.$$
Define $L_p^c = L_p^c(0, \infty)$ and $T_\rho^c : L_p^c \to L_q$ as in (4.11). Then the following holds.

PROPOSITION 5.5. *There exist bounded operators $X_i : l_p \to L_p^c$, $i = 0, 1, 2$, such that*
(5.7)
$$\|\mathcal{D}^i x\|_{l_q} \leq \|(T_\rho^c X_i)(x)\|_{L_q}$$
for all $x \in l_p$.

PROOF. We check (5.7) for $i = 0$; the remaining cases are treated similarly. If e_k, $k \in \mathbb{Z}$, denotes the k–th unit vector in l_p, the operator $X = X_0$ is uniquely defined by setting
$$X(e_k) := 2^{1/p'} |\Delta_{3k-1}|^{-1/p} \cdot \mathbf{1}_{\Delta_{3k-1}} - 2^{-1/p'} |\Delta_{3k+1}|^{-1/p} \cdot \mathbf{1}_{\Delta_{3k+1}}.$$
Of course, X is bounded as operator from l_p into L_p^c and the functions $g_k \in L_p^c$ defined with
$$g_k := (T_\rho^c X)(e_k)$$
possess the following properties:
$$\operatorname{supp}(g_k) \subseteq \Delta_{3k-1} \cup \Delta_{3k} \cup \Delta_{3k+1},$$
so they are disjointly supported and, moreover,
$$g_k(s) = 2^{1/p'} \cdot \rho(s) \cdot |\Delta_{3k-1}|^{1/p'} = |\Delta_{3k}|^{1/p'} \rho(s)$$

for $s \in \Delta_{3k}$. The values of g_k on the remaining two other intervals are not of interest later on. Consequently, given $x = (x_k)_{k\in\mathbb{Z}}$ in l_p, for $q < \infty$ it follows that

$$\begin{aligned}\left\|(T_\rho^c X)(x)\right\|_q^q &= \sum_{k\in\mathbb{Z}} |x_k|^q \cdot \|g_k\|_q^q \geq \sum_{k\in\mathbb{Z}} |x_k|^q \cdot \int_{\Delta_{3k}} g_k(s)^q\, ds \\ &= \sum_{k\in\mathbb{Z}} |x_k|^q \cdot |\Delta_{3k}|^{q/p'} \cdot \int_{\Delta_{3k}} \rho(s)^q\, ds \\ &= \sum_{k\in\mathbb{Z}} |x_k|^q \cdot \delta_{3k}^q = \left\|\mathcal{D}^0 x\right\|_q^q\end{aligned}$$

as claimed. For $q = \infty$ the similar estimate

$$\left\|(T_\rho^c X)(x)\right\|_\infty \geq \left\|\mathcal{D}^0 x\right\|_\infty$$

is trivial. $\square$

PROOF OF THEOREM 2.4 (1): In view of Lemma 4.2, inequality (5.7) leads to

$$e_n(\mathcal{D}^i) \leq e_n(T_\rho^c X_i) \leq \|X_i\|\, e_n(T_\rho^c) \leq c \cdot e_n(T_\rho)$$

(recall that T_ρ^c is the restriction of T_ρ to L_p^c, hence $e_n(T_\rho^c) \leq e_n(T_\rho)$) for $i = 0, 1, 2$. Thus,

(5.8) $$\sup_n n\, e_n(\mathcal{D}^i) \leq c \cdot \sup_n n\, e_n(T_\rho)$$

and it remains to estimate the left hand side of (5.8) from below. To do so we apply Proposition 1 of [**8**] to $(\delta_{3k+i})_{k\geq 0}$ and $(\delta_{3k+i})_{k<0}$ separately and obtain for the Lorentz norms of these sequences

(5.9) $$\|(\delta_{3k+i})_{k\geq 0}\|_{r,\infty} + \|(\delta_{3k+i})_{k<0}\|_{r,\infty} \leq c \cdot \sup_n n\, e_n(\mathcal{D}^i)$$

with some universal $c > 0$. Finally, since

$$\|(\delta_k)_{k\geq 0}\|_{r,\infty} \leq c \cdot \sum_{i=0}^{2} \|(\delta_{3k+i})_{k\geq 0}\|_{r,\infty}$$

(of course, a similar estimate holds for the sequences with $k < 0$) from (5.8) and (5.9) we derive

$$|\rho|_{r,\infty} = \|(\delta_k)_{k\geq 0}\|_{r,\infty} + \|(\delta_k)_{k<0}\|_{r,\infty} \leq c \cdot \sup_n n\, e_n(T_\rho)$$

as stated in (1) of Theorem 2.4. $\square$

PROOF OF THEOREM 2.4 (2): We have to verify

(5.10) $$\|\rho\|_r \leq c \cdot \liminf_{n\to\infty} n\, e_n(T_\rho)$$

with some universal $c > 0$. First note that it suffices to prove (5.10) for functions ρ with

$$\operatorname{supp}(\rho) \subseteq I = [K^{-1}, K]$$

for some positive K. Then $\rho \in L_q(I)$, and we are in the situation of Proposition 5.4. Applying this Proposition and using its notation, for $\varepsilon, \delta > 0$ we have

$$\begin{aligned}e_n(\lambda \cdot T_1 : F \to L_q(0,1)) &= e_n(Y \circ T_\rho \circ X : E \to L_q(I)) \\ &\leq \|Y\|\,\|X\| \cdot e_n(T_\rho : L_p(I) \to L_p(I)) \\ &\leq (1-\delta)^{-1} e_n(T_\rho : L_p(0,\infty) \to L_q(0,\infty))\,.\end{aligned}$$

Let $P : L_p(0,1) \to F$ be the projection with $\|P\| \leq 2$ (cf. Remark after Proposition 5.4). Then, if $m, n \in \mathbb{N}$, it follows that

$$e_{n+m-1}(T_1) \leq e_m(T_1 - T_1 P) + e_n(T_1 P),$$

and hence $\operatorname{rank}(T_1 - T_1 P) < \infty$ leads to

$$\liminf_{n \to \infty} n\, e_n(T_1) \leq \|P\| \liminf_{n \to \infty} n\, e_n(T_1 : F \to L_q(0,1)).$$

Consequently, we derive

$$\liminf_{n \to \infty} n\, e_n(T_\rho) \geq \lambda \cdot (1-\delta) \cdot \liminf_{n \to \infty} n\, e_n(T_1 : F \to L_q(0,1))$$
$$\geq (\|\rho\|_r^r - \varepsilon)^{1/r} \cdot (1-\delta) \cdot c,$$

with

$$c = (1/2) \liminf_{n \to \infty} n\, e_n(T_1 : L_p(0,1) \to L_q(0,1)) > 0$$

by Theorem 2.1. Letting $\varepsilon, \delta \to 0$ completes the proof of (5.10) in the case $q < \infty$.

In the case $q = \infty$ apply (5.10) to T_{ρ^*}. Recall $e_n(T_\rho) = e_n(T_{\rho^*})$, so we may replace $\|\rho\|_r$ by $\|\rho^*\|_r$ in the lower bound for $e_n(T_\rho)$. □

Our next objective is to prove part (3) of Theorem 2.4. Let us start with a quite general result which, roughly speaking, tells us the following. If we use, as before, the splitting $T_\rho = T_\rho^\circ P^\circ + T_\rho^c P^c$ with respect to the dyadic partition of $(0, \infty)$, then the operator T_ρ^c may become arbitrarily large with $\sup_n n\, e_n(T_\rho^\circ)$ bounded by a universal constant. More precisely, the following holds.

PROPOSITION 5.6. *Assume $1 < p \leq \infty$ and $1 \leq q \leq \infty$. Let $(d_k)_{k=1}^\infty$ be an arbitrary sequence of positive numbers. Then there exists a function $\rho \geq 0$ on $(0, \infty)$ such that*

(1) $\qquad d_k = \delta_k(\rho) = 2^{k/p'} \|\rho\|_{L_q(\Delta_k)}, \quad k = 1, 2, \ldots, \quad$ *and*

(2) $\qquad e_n(T_\rho^\circ : L_p^\circ(0, \infty) \to L_q(0, \infty)) \leq c\, n^{-1}$

for some universal $c > 0$.

PROOF. Let us choose intervals $I_k \subseteq \Delta_k$ with left–end point 2^k, i.e. the intervals are placed at the very left side of the Δ_k's. We choose the length of these intervals small enough, in order to provide for given d_k's

(5.11) $\qquad \displaystyle\sum_{k=1}^\infty 2^{-kr/p'} \cdot d_k^r \cdot |I_k|^{r/p'} \leq 1.$

Consequently, setting

(5.12) $\qquad \alpha_k := |I_k|^{-1/q} \cdot d_k \cdot 2^{-k/p'},$

by (5.11) we obtain

(5.13) $\qquad \displaystyle\sum_{k=1}^\infty \alpha_k^r \cdot |I_k| \leq 1.$

With these intervals I_k and α_k's we define a function ρ as

$$\rho := \sum_{k=1}^\infty \alpha_k \cdot \mathbf{1}_{I_k},$$

and hence by (5.13) we get $\rho \in L_r$ with $\|\rho\|_r \leq 1$. Now we verify that ρ possesses the asserted properties. Of course, since

$$2^{k/p'} \|\rho\|_{L_q(\Delta_k)} = 2^{k/p'} \cdot \alpha_k \cdot |I_k|^{1/q}$$

property (1) of the proposition is a direct consequence of (5.12).

To verify assertion (2), notice that we are in the situation described in the Remark after Proposition 4.1. We apply this Remark to the partition generated by $I_1, I_2, \ldots$ and $\Delta_1 \setminus I_1, \Delta_2 \setminus I_2, \ldots$. Thus for T_ρ^* defined in this Remark we have $T_\rho^* = T_\rho$ on $L_p^\circ(0, \infty)$; in particular, for all $f \in L_p^\circ(0, \infty)$ it follows that $T_\rho^* f(s) = T_\rho f(s) = 0$ provided that $s \in \Delta_k \setminus I_k$ for some $k \geq 1$. Thus (4.3) yields (those terms in (4.3), which are related to $\Delta_k \setminus I_k$, disappear)

$$\sup_n n \, e_n(T_\rho^\circ) \leq c \cdot \left\{ \sum_{k=1}^{\infty} \beta_k^r \, |I_k|^{r/p'} \right\}^{1/r} = c \cdot \left\{ \sum_{k=1}^{\infty} \alpha_k^r \, |I_k|^{r/p' + r/q} \right\}^{1/r} = c \cdot \|\rho\|_r \leq c \, .$$

This bound concludes the proof of the proposition. □

PROOF OF THEOREM 2.4 (3): If we apply Proposition 5.6 with $d_k = k^{-1/r}$, there exists a function ρ such that $\sup_n n \, e_n(T_\rho^\circ) < \infty$ and, moreover, $\delta_k(\rho) = k^{-1/r}$, $k \geq 1$. Hence $|\rho|_{r,\infty} = 1$, and by formula (4.14) of Proposition 4.3 we also have $\sup_n n \, e_n(T_\rho^c) < \infty$. Consequently, this also holds for T_ρ and completes the proof of Theorem 2.4 (3). □

5.3. Lower Bounds for Operators $T_{\rho,\psi}$ and $S_{\chi,\eta}$

Let us start with the generalization of Theorem 2.4 to operators $T_{\rho,\psi}$ mapping $L_p(0, \infty)$ to $L_q(0, \infty)$. Our first goal is to treat the case $p > 1$.

THEOREM 5.7. *Assume $1 < p \leq \infty$ and $1 \leq q \leq \infty$ and let ρ and ψ be as before.*

(1) *For $\bar{\rho} = \rho \cdot \mathbf{1}_{\{\psi > 0\}}$ and $\delta_k(\bar{\rho}, \psi)$ defined in (3.35) we have*

$$\left\| \left(\delta_k(\bar{\rho}, \psi) \right)_{k \in \mathbb{Z}} \right\|_{r,\infty} \leq c \cdot \sup_n n \, e_n(T_{\rho,\psi}) \, .$$

(2) *It also follows that*

$$\|\rho \, \psi\|_r \leq c \cdot \liminf_{n \to \infty} n \, e_n(T_{\rho,\psi})$$

with $\|\rho^* \, \psi\|_{p'}$ on the left hand side for $q = \infty$.

PROOF. We use the scale transformation (3.22). Let $\tilde{\rho}$ be defined in (3.24). In view of Proposition 3.11 both assertions follow directly by applying Proposition 3.2 and Theorem 2.4 to $\tilde{\rho}$, see (3.39) and (3.40).

For $q = \infty$ we may again replace $T_{\rho,\psi}$ by $T_{\rho^*,\psi}$ and apply Theorem 2.4 to $\widetilde{\rho^*}$. Note that $\|T_{\rho,\psi} f\|_\infty = \|T_{\rho^*,\psi} f\|_\infty$ for $f \in L_p$, so by the same arguments as for T_ρ we get $e_n(T_{\rho,\psi}) = e_n(T_{\rho^*,\psi})$ in this general situation as well. □

Remark: Assertion (1) may be rather weak for ψ's vanishing on "large" subsets since it does not take into account the contribution of the operator $T_{\rho-\bar{\rho},\psi}$.

Our next aim is to translate the estimates for $e_n(T_{\rho,\psi})$ in those for forward integration operators $S_{\chi,\eta} : L_p \to L_q$ defined in (3.21).

THEOREM 5.8. *Let χ and η be as before and regard the forward integration operator $S_{\chi,\eta} : L_p(0,\infty) \to L_q(0,\infty)$ with $p > 1$.*

(1) *For $\bar{\chi} = \chi \cdot \mathbf{1}_{\{\eta > 0\}}$ and $\delta_k^*(\bar{\chi}, \eta)$ defined in (3.35) we have*

$$\left\| \left(\delta_k^*(\bar{\chi}, \eta) \right)_{k \in \mathbb{Z}} \right\|_{r,\infty} \leq c \cdot \sup_n n \, e_n(S_{\chi,\eta}) \, .$$

(2) *It also follows that*

$$\|\chi \eta\|_r \leq c \cdot \liminf_{n \to \infty} n \, e_n(S_{\chi,\eta}) \, ,$$

with $\|\chi^* \eta\|_{p'}$ on the left hand side for $q = \infty$.

PROOF. We know from Proposition 3.7 that $S_{\chi,\eta}$ is isomorphic to $T_{\tilde{\chi},\tilde{\eta}}$ with

(5.14) $$\begin{aligned} \tilde{\chi}(\tau) &= \chi(\tau^{-1}) \cdot \tau^{-2/q} \quad \text{and} \\ \tilde{\eta}(\tau) &= \eta(\tau^{-1}) \cdot \tau^{-2/p'} \, . \end{aligned}$$

Since $S_{\chi,\eta}$ and $T_{\tilde{\chi},\tilde{\eta}}$ are isomorphic, they possess the same entropy behaviour. Both assertions follow directly by applying Theorem 5.7 to $T_{\tilde{\chi},\tilde{\eta}}$ with these $\tilde{\chi}$ and $\tilde{\eta}$, and by using Proposition 3.11 which yields $\delta_k^*(\chi, \eta) = \delta_k(\tilde{\chi}, \tilde{\eta})$ and $\|\chi \eta\|_r = \|\tilde{\chi} \tilde{\eta}\|_r$.

The case $q = \infty$ may be treated as before by starting with $S_{\chi^*,\eta}$ instead of $S_{\chi,\eta}$. Observe that the transformation (5.14) as well as its inverse are continuous. □

Now we can state lower bounds for the case $p = 1$ which are not covered by Theorem 5.7.

THEOREM 5.9. *Let $1 < q < \infty$, $\rho \in L_q(x, \infty)$, $\psi \in L_\infty(0, x)$, $x > 0$. Regarding $T_{\rho,\psi}$ as operator from $L_1(0,\infty)$ to $L_q(0,\infty)$ the following is valid.*

(1) *We have*

$$\left\| \left(\delta_k(\psi \mathbf{1}_{\{\rho > 0\}}, \rho) \right)_{k \in \mathbb{Z}} \right\|_{q,\infty} \leq c \cdot \sup_n n \, e_n(T_{\rho,\psi}) \, .$$

(2) *Suppose that the right hand sum of (4.44) in Theorem 4.9 is finite. Then we have*

(5.15) $$c_1 \|\rho \psi^*\|_q \leq \liminf_{n \to \infty} n \, e_n(T_{\rho,\psi}) \leq \limsup_{n \to \infty} n \, e_n(T_{\rho,\psi}) \leq c_2 \|\rho \psi^*\|_q \, .$$

PROOF. Assertion (1) follows directly from Proposition 4.11 for $T_{\rho,\psi}$ and its dual $S_{\psi,\rho}$, and by applying Theorem 5.8 to $S_{\psi,\rho}$.

The verification of (2) is a little bit more complicated. First note that the right hand estimate of (5.15) has already been proved, cf. (4.46) in Theorem 4.9. So let us prove the left hand one. An application of (4.62) to $T_{\rho,\psi} : L_1 \to L_q$ and its dual $S_{\psi,\rho} : L_{q'} \to L_\infty$ leads to

$$2n \, e_{2n}(S_{\psi,\rho}) \leq c \cdot (n \, e_n(S_{\psi,\rho}))^{\gamma/(\gamma+1)} \cdot (n \, e_n(T_{\rho,\psi}))^{1/(\gamma+1)}$$

with $c > 0$ and $\gamma \geq 2$ as in (4.62) only depending on q. Using Theorem 5.8 and (4.60) we derive from this

$$\begin{aligned} \|\rho \psi^*\|_q &\leq c \cdot \liminf_{n \to \infty} (2n) \, e_{2n}(S_{\psi,\rho}) \\ &\leq c \cdot \left(\limsup_{n \to \infty} n \, e_n(S_{\psi,\rho}) \right)^{\gamma/(\gamma+1)} \cdot \left(\liminf_{n \to \infty} n \, e_n(T_{\rho,\psi}) \right)^{1/(\gamma+1)} \\ &\leq c \cdot \|\rho \psi^*\|_q^{\gamma/(\gamma+1)} \cdot \left(\liminf_{n \to \infty} n \, e_n(T_{\rho,\psi}) \right)^{1/(\gamma+1)} \end{aligned}$$

proving part (2).

5.4. Regular Kernel Functions

Our next aim is to prove Corollary 2.5. We start with the following useful lemma which allows us to compare different norms of ρ.

LEMMA 5.10. *Let α be a real number, $r > 0$, and $k \in \mathbb{Z}$.*

(1) *If the function $\rho(t)\, t^\alpha$ is decreasing on $\Delta_{k-1} \cup \Delta_k$, then*

(5.16) $$\delta_k(\rho)^r \leq 2^{1+2|\alpha|r} \int_{\Delta_{k-1}} \rho(t)^r dt.$$

(2) *If the function $\rho(t)\, t^\alpha$ is increasing on $\Delta_k \cup \Delta_{k+1}$, then*

$$\delta_k(\rho)^r \leq 2^{1+2|\alpha|r} \int_{\Delta_{k+1}} \rho(t)^r dt.$$

PROOF. (1). Notice that for all $t' \in \Delta_{k-1}$ and $t \in \Delta_k$ we have

$$\rho(t) \leq \rho(2^k)\,(2^k/t)^\alpha \leq \rho(2^k)\,2^{|\alpha|},$$
$$\rho(t') \geq \rho(2^k)\,(2^k/t')^\alpha \geq \rho(2^k)\,2^{-|\alpha|}.$$

It follows that

$$\delta_k(\rho)^r = 2^{kr/p'} \|\rho\|^r_{L_q(\Delta_k)} \leq 2^{kr/p'} |\Delta_k|^{r/q} \rho(2^k)^r 2^{|\alpha|r} = 2^{k+|\alpha|r} \rho(2^k)^r$$

and

$$\int_{\Delta_{k-1}} \rho(t')^r dt' \geq |\Delta_{k-1}| \rho(2^k)^r 2^{-|\alpha|r} = 2^{k-1-|\alpha|r} \rho(2^k)^r.$$

Now (5.16) follows by comparing these two bounds. The proof of (2) is exactly the same. □

COROLLARY 5.11. *If the function $\rho(t)t^\alpha$ is monotone on $(0,\infty)$, then*

$$|\rho|_r \leq 2^{1/r+2|\alpha|}\,\|\rho\|_r.$$

COROLLARY 5.12. *Let α, β be two real numbers. If the function $\rho(t)\, t^\alpha$ is monotone in a neighborhood of zero and the function $\rho(t)\, t^\beta$ is monotone in a neighborhood of infinity, then $\|\rho\|_r < \infty$ yields $|\rho|_r < \infty$.*

PROOF OF COROLLARY 2.5: . The lower bound in (1) is proved in Theorem 2.4 (2). Furthermore, if $\|\rho\|_r < \infty$, it follows from Corollary 5.12 that $|\rho|_r < \infty$. The asserted upper bound in (1) is now a consequence of Theorem 2.2 (3).

Again the lower bound in (2) follows from Theorem 2.4 (2). To verify the upper bound, let $\rho \in L_q(0,\infty)$. Then

$$\sum_{k \leq 0} \delta_k(\rho)^r = \sum_{k \leq 0} 2^{kr/p'} \|\rho\|^r_{L_q(\Delta_k)} < \infty,$$

while the sum over positive k's is finite by Lemma 5.10. We thus obtain again that $|\rho|_r < \infty$, and hence the above mentioned argument works.

The upper bound of (3) follows from Corollary 5.11 combined with Theorem 2.2 (1). The lower bound was already proved in assertion (1). □

Our last aim is to adapt Corollary 2.5 to the case of two functions ρ and ψ. The key point here is to find conditions for ρ and ψ such that $\|\rho\,\psi\|_r < \infty$ necessarily implies $|(\rho\,\psi)|_r < \infty$ (cf. (3.36) for the definition). Note that $\|\rho\,\psi\|_r < \infty$ does not give any information about the size of ρ on the set $\{\psi = 0\}$, thus the above question makes only sense under the additional assumption $\bar{\rho} = \rho$, i.e. we assume

$$\rho = \rho \cdot \mathbf{1}_{\{\psi>0\}} \quad a.e..$$

Letting, as in Example 3.3(a),

$$\theta(s) := \int_0^s \psi(t)^{p'}\, dt$$

and

(5.17) $$\tilde{\rho}(\tau) = \rho(\theta^-\tau) \cdot \psi(\theta^-\tau)^{-p'/q}$$

we already know from Proposition 3.11 that $\|\tilde{\rho}\|_r = \|\rho\,\psi\|_r$, $\|\tilde{\rho}\|_q = \|\rho\|_q$ as well as

$$|\tilde{\rho}|_r = |(\bar{\rho},\psi)|_r = |(\rho,\psi)|_r\,.$$

Hence we obtain the following version of Corollary 2.5.

COROLLARY 5.13. Suppose $p > 1$ and $\rho = \rho \cdot \mathbf{1}_{\{\psi>0\}}$ a.e.. Let $\tilde{\rho}$ be defined in (5.17).

(1) Assume that for some real numbers α, β the functions $\tilde{\rho}(\tau)\,\tau^\alpha$ and $\tilde{\rho}(\tau)\,\tau^\beta$ are monotone in a neighborhood of zero and infinity, respectively. Then

(5.18) $$c_1\,\|\rho\,\psi\|_r \leq \liminf_{n\to\infty} n\, e_n(T_{\rho,\psi}) \leq \limsup_{n\to\infty} n\, e_n(T_{\rho,\psi}) \leq c_2\,\|\rho\,\psi\|_r\,.$$

(2) In the previous statement one may replace the assumption about monotonicity at zero by supposing $\tilde{\rho} \in L_q(0,\infty)$.

(3) Assume that for some real number α the function $\tilde{\rho}(\tau)\,\tau^\alpha$ is m   e on $(0,\infty)$. Then

(5.19) $$c_1\,\|\rho\,\psi\|_r \leq \sup_n n\, e_n(T_{\rho,\psi}) \leq c_2\, c_\alpha\,\|\rho\,\psi\|_r$$

with $c_\alpha = 2^{1/r+2|\alpha|}$.

(4) For $q = \infty$ we have to replace $\|\rho\,\psi\|_r$ by $\|\rho^*\,\psi\|_r = \|\rho^*\,\psi\|_{p'}$ in (5.18) and (5.19).

Remark: If the function

(5.20) $$t \to \frac{\rho(t)}{\psi(t)^{p'/q}}$$

is monotone, then $\tilde{\rho}$ is also monotone. Hence, when (5.20) is monotone in neighborhoods of zero and infinity (or everywhere), assumptions (1) or (3) of Corollary 5.13 are satisfied and we obtain (5.18) or (5.19), respectively.

In the case $p = 1$, not covered by Corollary 5.13, the following are true.

COROLLARY 5.14. Let $T_{\rho,\psi} : L_1(0,\infty) \to L_q(0,\infty)$ with $1 < q < \infty$ and assume $\psi = \psi \cdot \mathbf{1}_{\{\rho>0\}}$ a.e..

(1) If ψ is monotone near zero and infinity, then

$$c_1\,\|\rho\,\psi^*\|_q \leq \liminf_{n\to\infty} n\, e_n(T_{\rho,\psi}) \leq \limsup_{n\to\infty} n\, e_n(T_{\rho,\psi}) \leq c_2\,\|\rho\,\psi^*\|_q\,.$$

(2) If ψ is monotone on $(0,\infty)$, then
$$c_1 \left\|\rho\,\psi^*\right\|_q \leq \sup_n n\, e_n(T_{\rho,\psi}) \leq c_2 \left\|\rho\,\psi^*\right\|_q .$$

PROOF. For $T_{\rho,\psi} : L_1(0,\infty) \to L_q(0,\infty)$ we regard, as before, its dual $S_{\psi,\rho} : L_{q'}(0,\infty) \to L_\infty(0,\infty)$ which is isomorphic to $T_{\tilde\psi,\tilde\rho}$ with

(5.21) $\qquad \tilde\psi(\tau) = \psi(\tau^{-1}) \quad\text{and}\quad \tilde\rho(\tau) = \rho(\tau^{-1})\cdot \tau^{-2/q}, \quad \tau > 0 .$

An application of the preceding remark to $T_{\tilde\psi,\tilde\rho}$ (as a mapping from $L_{q'}$ into L_∞) implies the following: If $\tilde\psi$ is monotone near zero and infinity (of course, this is equivalent to ψ being monotone), part (4) of Corollary 5.13 lets us conclude

(5.22) $\qquad c_1 \left\|(\tilde\psi)^* \tilde\rho\right\|_q \leq \liminf_{n\to\infty} n\, e_n(T_{\tilde\psi,\tilde\rho}) \leq \limsup_{n\to\infty} n\, e_n(T_{\tilde\psi,\tilde\rho}) \leq c_2 \left\|(\tilde\psi)^* \tilde\rho\right\|_q .$

As before, $(\tilde\psi)^* = \widetilde{\psi^*}$, hence by Proposition 3.11 we derive from (5.22)

(5.23) $\qquad c_1 \left\|\rho\,\psi^*\right\|_q \leq \liminf_{n\to\infty} n\, e_n(S_{\psi,\rho}) \leq \limsup_{n\to\infty} n\, e_n(S_{\psi,\psi}) \leq c_2 \left\|\rho\,\psi^*\right\|_q .$

As always, $c_1, c_2 > 0$ are universal, only depending on q. Hence we may apply (5.15) (more precisely, the proof of (5.15)) and get

$$c_1 \left\|\rho\,\psi^*\right\|_q \leq \liminf_{n\to\infty} n\, e_n(T_{\rho,\psi}) \leq \limsup_{n\to\infty} n\, e_n(T_{\rho,\psi}) \leq c_2 \left\|\rho\,\psi^*\right\|_q$$

as claimed in (1).

The second assertion follows exactly along the same lines by using Theorem 4.11 (instead of (5.15)) in the last step. $\square$

CHAPTER 6

Approximation Numbers

The aim of this chapter is to prove optimal estimates for $a_n(T_\rho)$ or $a_n(T_{\rho,\psi})$ based on integrability properties of ρ or ρ and ψ, respectively. Thereby T_ρ and $T_{\rho,\psi}$ are regarded as operators from $L_p(0,\infty)$ into $L_q(0,\infty)$ with $1 \leq p,q \leq \infty$. This extends former results of Edmunds et al. obtained for the case $1 < p = q < \infty$ ([**17**], [**18**]) as well as of Evans et al. for $p = q = 1$ and $p = q = \infty$ ([**20**]).

6.1. Maz'ja–Rosin Theorem

We start with a version of the well-known Maz'ja–Rosin Theorem (cf. [**37**], p. 39–51 or [**41**] and the references stated there). Recall that this theorem describes σ–finite measures ν and μ on $\mathbb{R}^+$ for which

(6.1) $$T_1 : L_p(\mathbb{R}^+, \nu) \to L_q(\mathbb{R}^+, \mu)$$

is bounded. If $\psi > 0$ a.e., this easily yields conditions for the boundedness of

$$T_{\rho,\psi} : L_p(\mathbb{R}^+, dt) \to L_q(\mathbb{R}^+, ds) ,$$

yet for general ψ's additional considerations are necessary.

THEOREM 6.1. *Let $1 \leq p,q \leq \infty$ and $\rho, \psi \geq 0$ on $(0,\infty)$. Then $T_{\rho,\psi}$ is bounded from $L_p(0,\infty)$ into $L_q(0,\infty)$ iff*

$$B_{p,q} = B_{p,q}(\rho, \psi) < \infty ,$$

where

$$B_{p,q} = \sup_{s>0} \|\psi\|_{L_{p'}(0,s)} \cdot \|\rho\|_{L_q(s,\infty)}$$

if $1 \leq p \leq q \leq \infty$ and

$$B_{p,q} = \left(\int_0^\infty \left[\|\psi\|_{L_{p'}(0,s)}^{p'(q-1)} \cdot \|\rho\|_{L_q(s,\infty)}^q \right]^{p/(p-q)} \cdot \psi(s)^{p'} ds \right)^{(p-q)/pq}$$

for $1 \leq q < p \leq \infty$.
Moreover, there are universal constants $c_{p,q}, C_{p,q} > 0$ such that

(6.2) $$c_{p,q} \cdot B_{p,q}(\rho, \psi) \leq \|T_{\rho,\psi} : L_p \to L_q\| \leq C_{p,q} \cdot B_{p,q}(\rho, \psi) .$$

PROOF. For $p = 1$ the assertion follows exactly as in the classical case apart from some minor modifications (cf. [**37**], p. 42-43). Similarly, for $q = \infty$, one may modify the classical proof.[1] Hence we restrict our consideration to $p > 1$ and $q < \infty$. As in Example 3.3 (a) we define the transformation θ by

$$\theta(s) = \int_0^s \psi(t)^{p'} dt ,$$

---

[1] Another way could be via duality arguments and Proposition 3.8.

and borrow the related notation from section 3. The operator $T_{\rho,\psi}$ is bounded iff $T_{\rho,\psi} \circ \Phi_p^\theta$ is so from $L_p(A, d\tau)$ into $L_q(0,\infty)$ with $A = (0, \theta(\infty))$. Now define a measure μ on A by

$$(6.3) \qquad \mu(B) := \int_{\{\theta s \in B\}} \rho(s)^q\, ds$$

for $B \subseteq A$ measurable. Then for $g \in L_p(A, d\tau)$ by Proposition 3.2 we get

$$\|(T_{\rho,\psi} \circ \Phi_p^\theta)g\|_q^q = \int_0^\infty \left|\int_0^{\theta(s)} g(\tau)\, d\tau\right|^q ds$$

$$= \int_0^\infty \left|\int_0^\sigma g(\tau)\, d\tau\right|^q d\mu(\sigma),$$

i.e. the operator $T_{\rho,\psi}$ is bounded iff this is so for T_1 acting from $L_p(A, d\tau)$ into $L_q(A, \mu)$. Hence we are in the classical situation (6.1) with Lebesgue measure ν and μ defined by (6.3). Moreover, since Φ_p^θ is isometric,

$$(6.4) \qquad \|T_{\rho,\psi} : L_p(0,\infty) \to L_q(0,\infty)\| = \|T_1 : L_p(A, d\tau) \to L_q(A, \mu)\|.$$

Consequently, if $p \le q$ (the case $q < p$ may be treated similarly), the mapping $T_{\rho,\psi}$ is bounded from $L_p(0,\infty)$ into $L_q(0,\infty)$ iff

$$(6.5) \qquad \sup_{0 < \sigma < \theta(\infty)} \sigma^{1/p'} \cdot \mu[\sigma, \theta(\infty))^{1/q} < \infty.$$

By the definition of μ we have

$$\mu[\sigma, \theta(\infty)) = \int_{\{\theta t \ge \sigma\}} \rho(t)^q\, dt = \int_{\theta^-\sigma}^\infty \rho(t)^q\, dt,$$

thus (6.5) is equivalent to

$$(6.6) \qquad \sup_{0 < s < \infty} \theta(s)^{1/p'} \cdot \left(\int_{\theta^-\theta s}^\infty \rho(t)^q\, dt\right)^{1/q} < \infty.$$

It is easy to see that the expression in (6.6) equals

$$(6.7) \qquad \sup_{0 < s < \infty} \theta(s)^{1/p'} \cdot \|\rho\|_{L_q(s,\infty)} = \sup_{0 < s < \infty} \|\psi\|_{L_{p'}(0,s)} \cdot \|\rho\|_{L_q(s,\infty)}.$$

Consequently, the condition $B_{p,q}(\rho, \psi) < \infty$ is necessary and sufficient for the boundedness of $T_{\rho,\psi}$. In view of (6.4) and (6.7), the estimates in (6.2) follow by the classical estimates (see [37]) for $\|T_1 : L_p(\mathbb{R}^+, \nu) \to L_q(\mathbb{R}^+, \mu)\|$. □

Remark: The constant $C_{p,q}$ appearing in (6.2) may be chosen as

$$\begin{array}{rcll} C_{p,q} &=& (1/q')^{1/p'} \cdot q^{1/q} &: 1 < p \le q < \infty, \\ C_{p,q} &=& 1 &: p = 1 \text{ or } q = 1 \text{ or } q = \infty, \\ C_{p,q} &=& (1/p')^{1/q'} \cdot q^{1/q} &: 1 < q < p \le \infty. \end{array}$$

COROLLARY 6.2. *For any finite interval $I \subset (0,\infty)$ and $\rho \in L_q(I)$ it follows that*

$$\|T_\rho f\|_q \le C_{p,q} \cdot |I|^{1/p'} \cdot \|\rho\|_{L_q(I)} \|f\|_p$$

for all $f \in L_p(I)$.

PROOF. For $1 \leq p \leq q \leq \infty$ the constant $B_{p,q}$ appearing in Theorem 6.1 can be estimated as
$$B_{p,q}(\rho, 1) = \sup_{s \in I} \|\mathbf{1}_{I \cap (0,s)}\|_{p'} \cdot \|\rho \mathbf{1}_{(s,\infty)}\|_q \leq |I|^{1/p'} \cdot \|\rho\|_q \ ,$$
while for $1 \leq q < p \leq \infty$ by the same argument
$$\begin{aligned} B_{p,q}(\rho, 1) &\leq |I|^{[(q-1)\frac{p}{p-q}+1]\frac{p-q}{pq}} \cdot \|\rho\|_q \\ &= |I|^{1-1/q+1/q-1/p} \cdot \|\rho\|_q = |I|^{1/p'} \cdot \|\rho\|_q \end{aligned}$$
completing the proof. $\square$

6.2. Estimates for $a_n(T_\rho)$ on Finite Intervals

Our aim is to find optimal estimates for $a_n(T_\rho)$ with $T_\rho : L_p(I) \to L_q(I)$. Thereby the quantity
$$J_\rho(I) := \|\mathbf{1}\|_{L_{p'}(I)} \cdot \|\rho\|_{L_q(I)}$$
will play an important role. Note that we already used an expression defined in this way, namely, with this notation (2.9) can now be written as
$$|\rho|_r = \left(\sum_{k=-\infty}^{\infty} J_\rho(\Delta_k)^r \right)^{1/r}$$
with dyadic intervals Δ_k defined in (2.7).

We start with proving some easy properties of J_ρ for later use. As before, $1/r = 1/p' + 1/q > 0$.

LEMMA 6.3. (1) Let $\{I_1, \ldots, I_N\}$ be an arbitrary partition of interval I. Then we have

(6.8)
$$\left(\sum_{k=1}^{N} J_\rho(I_k)^r \right)^{1/r} \leq J_\rho(I) \ .$$

(2) For $\rho_1, \rho_2 \in L_q(I)$ and each partition $\{I_1, \ldots, I_N\}$ of I it follows that

(6.9)
$$\left| \left(\sum_{k=1}^{N} J_{\rho_1}(I_k)^r \right)^{1/r} - \left(\sum_{k=1}^{N} J_{\rho_2}(I_k)^r \right)^{1/r} \right| \leq J_{|\rho_1 - \rho_2|}(I)$$

provided that $1 \leq r < \infty$. For $0 < r \leq 1$ we have

(6.10)
$$\left| \sum_{k=1}^{N} J_{\rho_1}(I_k)^r - \sum_{k=1}^{N} J_{\rho_2}(I_k)^r \right| \leq J_{|\rho_1 - \rho_2|}(I)^r \ .$$

PROOF. Hölder's inequality gives
$$\begin{aligned} \sum_{k=1}^{N} |I_k|^{r/p'} \cdot \left(\int_{I_k} \rho(s)^q \, ds \right)^{r/q} &\leq \left(\sum_{k=1}^{N} |I_k| \right)^{r/p'} \cdot \left(\sum_{k=1}^{N} \int_{I_k} \rho(s)^q \, ds \right)^{r/q} \\ &= \left| \bigcup_{k=1}^{N} I_k \right|^{r/p'} \cdot \|\rho\|_{L_q(\bigcup_1^N I_k)}^r = J_\rho \left(\bigcup_{k=1}^{N} I_k \right)^r , \end{aligned}$$

proving (6.8).

To verify (6.9), observe that for any interval $A \subseteq I$

$$|J_{\rho_1}(A) - J_{\rho_2}(A)| \leq |A|^{1/p'} \cdot \|\rho_1 - \rho_2\|_{L_q(A)} = J_{|\rho_1-\rho_2|}(A) \,.$$

Thus, if $r \geq 1$ this implies

$$\left|\left(\sum_{k=1}^{N} J_{\rho_1}(I_k)^r\right)^{1/r} - \left(\sum_{k=1}^{N} J_{\rho_2}(I_k)^r\right)^{1/r}\right|$$
$$\leq \left(\sum_{k=1}^{N} |J_{\rho_1}(I_k) - J_{\rho_2}(I_k)|^r\right)^{1/r} \leq \left(\sum_{k=1}^{N} J_{|\rho_1-\rho_2|}(I_k)^r\right)^{1/r} \leq J_{|\rho_1-\rho_2|}(I) \,,$$

where the last estimate follows by (6.8).

The case $0 < r < 1$ can be proved in the same way by using the inequality

$$|\|x\|_r^r - \|y\|_r^r| \leq \|x-y\|_r^r$$

for any two vectors $x, y \in \mathbb{R}^N$ and the usual l_r–quasinorm on $\mathbb{R}^N$. □

The next two properties are crucial for estimating the approximation numbers of T_ρ.

PROPOSITION 6.4. *Let $I \subset (0,\infty)$ be a finite interval and let $\rho \in L_q(I)$.*

(1) *If $q < \infty$, then*

$$(6.11) \quad \|\rho\|_{L_r(I)} = \inf_{\mathcal{I}} \left\{\left(\sum_{k=1}^{N} J_\rho(I_k)^r\right)^{1/r} : \mathcal{I} = \{I_1,\ldots,I_N\} \text{ partition of } I\right\} \,.$$

For $q = \infty$, equality (6.11) holds with $\|\rho^*\|_{L_r(I)}$ on the left hand side (recall that ρ^* was defined in (2.11)).

(2) *For each $n \in \mathbb{N}$ there exists a partition $\mathcal{I}^* = \{I_1^*,\ldots,I_n^*\}$ of I with*

$$(6.12) \quad J_\rho(I_1^*) = \cdots = J_\rho(I_n^*) \,.$$

PROOF. We split the proof of (6.11) into three steps. Our first aim is to show that the left hand side of (6.11) is always less or equal the right hand infimum. Indeed, if $\mathcal{I} = \{I_1,\ldots,I_N\}$ is an arbitrary partition of I, Hölder's inequality gives

$$\|\rho\|_r^r = \sum_{k=1}^{N} \int_{I_k} \rho(s)^r \, ds \leq \sum_{k=1}^{N} |I_k|^{r/p'} \cdot \left(\int_{I_k} \rho(s)^q \, ds\right)^{r/q}$$
$$= \sum_{k=1}^{N} J_\rho(I_k)^r$$

with obvious modifications for $p = 1$. This proves our first claim.

In a second step we verify (6.11) for i.s.f. φ, i.e. for functions φ of the form

$$\varphi = \sum_{k=1}^{N} \alpha_k \mathbf{1}_{I_k}$$

for some partition $I_1,\ldots,I_N$ of I and α_k's non–negative. Direct calculations give

$$(6.13) \quad \sum_{k=1}^{N} J_\varphi(I_k)^r = \sum_{k=1}^{N} \alpha_k^r \, |I_k|^{r/p'} \, |I_k|^{r/q} = \|\varphi\|_r^r \,,$$

so (6.11) holds for those functions φ by the first step of our proof.

In a last step we first treat the case $q < \infty$. Given $\rho \in L_q(I)$ and $\varepsilon > 0$ we choose an i.s.f. $\varphi \geq 0$ with

(6.14) $$\|\rho - \varphi\|_q \leq \varepsilon \cdot |I|^{-1/p'},$$

which clearly implies

(6.15) $$\|\rho - \varphi\|_r \leq \varepsilon.$$

Let $\mathcal{I} = \{I_1, \ldots, I_N\}$ be the partition generated by φ, so that by (6.13) we have

$$\left(\sum_{k=1}^N J_\varphi(I_k)^r\right)^{1/r} = \|\varphi\|_r.$$

Next assume $r \geq 1$. An application of (6.9), (6.14) and (6.15) leads to

$$\left(\sum_{k=1}^N J_\rho(I_k)^r\right)^{1/r} \leq \left(\sum_{k=1}^N J_\varphi(I_k)^r\right)^{1/r} + \varepsilon$$
$$= \|\varphi\|_r + \varepsilon \leq \|\rho\|_r + 2\varepsilon.$$

Since $\varepsilon > 0$ was arbitrary, this completes the proof for $r \geq 1$.

If $0 < r < 1$, in similar way we get by (6.10)

$$\sum_{k=1}^N J_\rho(I_k)^r \leq \sum_{k=1}^N J_\varphi(I_k)^r + \varepsilon^r = \|\varphi\|_r^r + \varepsilon^r \leq \|\rho\|_r^r + 2\varepsilon^r$$

which, of course, also proves (6.11) for those r's.

Thus it remains to treat the case $q = \infty$ where $r = p'$. In one direction, by the definition of ρ^* (see (2.11)) we have for each partition $\mathcal{I}$,

$$\|\rho^*\|_r^r = \sum_{k=1}^N \int_{I_k} \rho^*(s)^r \, ds \leq$$

$$\sum_{k=1}^N |I_k| \cdot \|\rho^*\|_{L_\infty(I_k)}^r \leq \sum_{k=1}^N |I_k| \cdot \|\rho\|_{L_\infty(I_k)}^r = \sum_{k=1}^N J_\rho(I_k)^r.$$

In the other direction, we fix an $\varepsilon > 0$ and using Lemma 4.5 we find an i.s.f. $\varphi \geq 0$ such that $\rho \leq \varphi$ a.e. but

$$\int_I \varphi(s)^r \, ds \leq \int_I \rho^*(s)^r \, ds + \varepsilon.$$

Let $\mathcal{I}$ be a partition associated with φ. Then, by the choice of φ,

$$\sum_{k=1}^N J_\rho(I_k)^r \leq \sum_{k=1}^N J_\varphi(I_k)^r = \|\varphi\|_r^r \leq \|\rho^*\|_r^r + \varepsilon,$$

and we prove again (6.11) by letting $\varepsilon \to 0$.

To verify (6.12), fix $n \in \mathbb{N}$ and define a function Φ on the set of n–partitions of I as follows:

$$\Phi(\mathcal{I}) := \max_{1 \leq k \leq n} J_\rho(I_k) - \min_{1 \leq k \leq n} J_\rho(I_k), \quad \mathcal{I} = \{I_1, \ldots, I_n\}.$$

Of course, $\Phi(\mathcal{I})$ depends continuously on the $n-1$ points in I determining $\mathcal{I}$, so that there exists a partition $\mathcal{I}^*$ where Φ attains its minimum. At this point we cannot exclude that one or more sets of $\mathcal{I}^*$ are degenerated. But we claim that

$$\Phi(\mathcal{I}^*) = 0, \tag{6.16}$$

which, in particular, implies $|I_k^*| > 0$ for all k's provided that $J_\rho(I) \neq 0$. Otherwise (6.12) trivially holds. To prove (6.16), let $\mathcal{I} = \{I_1, \ldots, I_n\}$ be an arbitrary n-partition of I with $\Phi(\mathcal{I}) > 0$. Then we find a $k_0 \in \{1, \ldots, n\}$ where $J_\rho(I_{k_0})$ is maximal, yet at least for one neighbor, say for $k_0 + 1$, we have

$$J_\rho(I_{k_0+1}) < J_\rho(I_{k_0}). \tag{6.17}$$

Now we move the right hand border of I_{k_0} to the left such that $J_\rho(I_{k_0})$ becomes strictly less with (6.17) still valid. In this way we obtain a new partition $\mathcal{I}'$ with $\Phi(\mathcal{I}') \leq \Phi(\mathcal{I})$ where the number of sets in $\mathcal{I}'$ with J_ρ maximal is strictly smaller than in $\mathcal{I}$. If we repeat this procedure finitely often, we finally arrive at a partition $\mathcal{I}''$ with $\Phi(\mathcal{I}'') < \Phi(\mathcal{I})$. In other words, any partition $\mathcal{I}$ with $\Phi(\mathcal{I}) > 0$ cannot be minimal and this proves (6.16). By the construction of Φ this completes the proof. $\square$

Before proceeding further, we recall and slightly modify some former notation. Let $I \subseteq (0, \infty)$ be an arbitrary interval and let $\mathcal{I} = \{I_k\}_{k \in \mathcal{K}}$ be a partition of I. Here $\mathcal{K}$ denotes either a finite or countable infinite index set and, furthermore, we always assume that the I_k's are bounded intervals with $\rho \in L_q(I_k)$ for each $k \in \mathcal{K}$. As in (3.31) we now set

$$L_p^\circ(I) = L_p^\circ(I; \mathcal{I}) := \left\{ f \in L_p(I) : \int_{I_k} f(t)\, dt = 0, \ k \in \mathcal{K} \right\}.$$

The complementary space

$$L_p^c(I) = L_p^c(I; \mathcal{I})$$

consists as before of functions in $L_p(I)$ which are constant on each interval I_k, $k \in \mathcal{K}$. The projections P° and P^c denote as above the corresponding (bounded) projections from $L_p(I)$ onto $L_p^\circ(I)$ and $L_p^c(I)$, respectively. Recall that $\|P^c\| \leq 1$, hence $\|P^\circ\| \leq 2$.

The next lemma is the key for all further estimates.

LEMMA 6.5. *Let $I \subseteq (0, \infty)$ be some interval (not necessarily finite) and let $\mathcal{I} = \{I_k\}_{k \in \mathcal{K}}$ be a partition of I. Then for every $f \in L_p^\circ(I; \mathcal{I})$ we have*

$$\|T_\rho f\|_q \leq \begin{cases} C_{p,q} \sup_{k \in \mathcal{K}} J_\rho(I_k) \|f\|_p & : 1 \leq p \leq q \leq \infty, \\ C_{p,q} \left(\sum_{k \in \mathcal{K}} J_\rho(I_k)^{r/(1-r)} \right)^{1/r - 1} \|f\|_p & : 1 \leq q < p \leq \infty. \end{cases}$$

PROOF. Notice that each function $T_\rho(f\, \mathbf{1}_{I_k})$ is supported by its own interval I_k. Hence Corollary 6.2 implies

$$\|T_\rho f\|_q^q = \sum_{k \in \mathcal{K}} \|T_\rho(f\, \mathbf{1}_{I_k})\|_q^q \leq C_{p,q}^q \sum_{k \in \mathcal{K}} J_\rho(I_k)^q \|f\, \mathbf{1}_{I_k}\|_p^q.$$

For $p \leq q$ we replace all $J_\rho(I_k)$ by the maximal value and use the elementary estimate (monotinicity of the l_p-norms)
$$\sum_{k \in \mathcal{K}} \|f \mathbf{1}_{I_k}\|_p^q \leq \|f\|_p^q,$$
while for $q < p$ Hölder's inequality gives
$$\sum_{k \in \mathcal{K}} J_\rho(I_k)^q \|f \mathbf{1}_{I_k}\|_p^q \leq \left(\sum_{k \in \mathcal{K}} J_\rho(I_k)^{r/(1-r)}\right)^{q(1-r)/r} \left(\sum_{k \in \mathcal{K}} \|f \mathbf{1}_{I_k}\|_p^p\right)^{q/p}$$
$$= \left(\sum_{k \in \mathcal{K}} J_\rho(I_k)^{r/(1-r)}\right)^{q(1-r)/r} \|f\|_p^q,$$
and we are done in both cases. $\square$

Now we are in position to state and prove a first estimate for the approximation numbers.

THEOREM 6.6. *Let $I \subset (0, \infty)$ and $\mathcal{I} = \{I_k\}_{k \in \mathcal{K}}$ be as before. Denote as before by P° the projection from $L_p(I)$ onto $L_p^\circ(I; \mathcal{I})$. Given natural numbers $(n_k)_{k \in \mathcal{K}}$ with*

(6.18) $$n := \sum_{k \in \mathcal{K}} (n_k - 1) + 1 < \infty,$$

then it follows that
$$a_n(T_\rho P^\circ) \leq \begin{cases} 2\, C_{p,q} \sup_{k \in \mathcal{K}} n_k^{-1/r} J_\rho(I_k) & : \ 1 \leq p \leq q \leq \infty, \\ 2\, C_{p,q} \left(\sum_{k \in \mathcal{K}} n_k^{-r/(1-r)} J_\rho(I_k)^{r/(1-r)}\right)^{1/r - 1} & : \ 1 \leq q < p \leq \infty. \end{cases}$$

PROOF. First note that in view of (6.18) at most finitely many of the n_k's can be different from 1. For those n_k's we split the corresponding intervals I_k according to Proposition 6.4 into n_k intervals $I_{k,1}^*, \ldots, I_{k,n_k}^*$ with
$$J_\rho(I_{k,1}^*) = \cdots = J_\rho(I_{k,n_k}^*).$$
If $n_k = 1$, set $I_{k,1}^* := I_k$, so that by (6.8) for all $k \in \mathcal{K}$ and $1 \leq j \leq n_k$

(6.19) $$J_\rho(I_{k,j}^*) \leq n_k^{-1/r} \cdot J_\rho(I_k).$$

Now let
$$\mathcal{I}^* := \{I_{k,j}^* : 1 \leq j \leq n_k,\ k \in \mathcal{K}\}$$
be the partition of I generated by partitioning each I_k into n_k intervals $I_{k,j}^*$ for $1 \leq j \leq n_k$. If
$$P_*^c : L_p(I) \to L_p^c(I; \mathcal{I}^*) \quad \text{and} \quad P_*^\circ : L_p(I) \to L_p^\circ(I; \mathcal{I}^*)$$
denote the canonical projections with respect to the finer partition, then with $P^\circ : L_p(I) \to L_p(I; \mathcal{I})$ we have

(6.20) $$P^\circ - P_*^c P^\circ = P_*^\circ P^\circ = P_*^\circ$$

and, furthermore,

(6.21) $$\operatorname{rank}(P_*^c P^\circ) \leq \sum_{k \in \mathcal{K}} (n_k - 1) = n - 1.$$

If $p \leq q$, in view of (6.20), Lemma 6.5 and (6.19), for each $f \in L_p(I)$ we derive

$$\begin{aligned} \|(T_\rho P^\circ - T_\rho P_*^c P^\circ)f\|_q &= \|T_\rho P_*^\circ f\|_q \\ &\leq C_{p,q} \cdot [\sup_{k \in \mathcal{K}} \max_{j \leq n_k} J_\rho(I_{k,j}^*)] \cdot \|P_*^\circ f\|_p \\ &\leq 2\, C_{p,q} \cdot [\sup_{k \in \mathcal{K}} n_k^{-1/r} J_\rho(I_k)] \cdot \|f\|_p \ . \end{aligned}$$
(6.22)

For $q < p$ we obtain by similar arguments

$$\begin{aligned} \|(T_\rho P^\circ - T_\rho P_*^c P^\circ)f\|_q &\leq C_{p,q} \cdot \left(\sum_{k \in \mathcal{K}} \sum_{j \leq n_k} J_\rho(I_{k,j}^*)^{r/(1-r)} \right)^{1/r - 1} \cdot \|P_*^\circ f\|_p \\ &\leq 2\, C_{p,q} \cdot \left(\sum_{k \in \mathcal{K}} n_k^{1 - 1/(1-r)} J_\rho(I_k)^{r/(1-r)} \right)^{1/r - 1} \cdot \|f\|_p \\ &= 2\, C_{p,q} \cdot \left(\sum_{k \in \mathcal{K}} n_k^{-r/(1-r)} J_\rho(I_k)^{r/(1-r)} \right)^{1/r - 1} \cdot \|f\|_p \end{aligned}$$
(6.23)

for all $f \in L_p(I)$. By (6.21) it follows that

$$\mathrm{rank}(T_\rho P_*^c P^\circ) \leq \mathrm{rank}(P_*^c P^\circ) \leq n - 1 \ ,$$

so estimates (6.22) and (6.23) complete the proof by the definition of approximation numbers. $\square$

Remark: In the specific case $p = 2$ the operator P_*° is an orthogonal projector. Therefore its norm equals one and we can eliminate the factor 2 in the statement of Theorem 6.6 as well as later on in Corollary 6.7 and in Theorem 6.8.

COROLLARY 6.7. *For I and $\mathcal{I} = \{I_k\}_{k \in \mathcal{K}}$ as before and each $n \geq 1$ we have*

$$a_n(T_\rho P^\circ) \leq 2\, C_{p,q} \cdot n^{-\lambda} \cdot \left(\sum_{k \in \mathcal{K}} J_\rho(I_k)^r \right)^{1/r} ,$$
(6.24)

where

$$\lambda := \min\{1, 1/r\} \ .$$
(6.25)

PROOF. Of course we may suppose $\sum_{j \in \mathcal{K}} J_\rho(I_j)^r < \infty$. Otherwise (6.24) clearly holds. For $k \in \mathcal{K}$ choose integers $n_k \in \mathbb{N}$ satisfying

$$n_k - 1 < n \cdot \frac{J_\rho(I_k)^r}{\sum_{j \in \mathcal{K}} J_\rho(I_j)^r} \leq n_k \ .$$
(6.26)

Assume first $p \leq q$. By the choice of the n_k's we have $\sum_{k \in \mathcal{K}}(n_k - 1) \leq n - 1$. Thus Theorem 6.6 and (6.26) imply

$$\begin{aligned} a_n(T_\rho P^\circ) &\leq 2\, C_{p,q} \cdot [\sup_{k \in \mathcal{K}} n_k^{-1/r} J_\rho(I_k)] \\ &\leq 2\, C_{p,q} \cdot n^{-1/r} \cdot \left(\sum_{k \in \mathcal{K}} J_\rho(I_k)^r \right)^{1/r} \end{aligned}$$

as asserted.

For $q < p$ and each $k \in \mathcal{K}$ the right hand estimate of (6.26) yields

$$n_k^{-r/(1-r)} J_\rho(I_k)^{r/(1-r)} \leq n^{-r/(1-r)} \cdot \left(\sum_{j \in \mathcal{K}} J_\rho(I_j)^r\right)^{r/(1-r)} J_\rho(I_k)^r .$$

Consequently, this time Theorem 6.6 leads to

$$a_n(T_\rho P^\circ) \leq 2\, C_{p,q} \cdot n^{-1} \cdot \sum_{k \in \mathcal{K}} J_\rho(I_k)^r \cdot \left(\sum_{k \in \mathcal{K}} J_\rho(I_k)^r\right)^{1/r - 1}$$

which, of course, completes the proof in this case as well. □

THEOREM 6.8. *Let $I \subset (0, \infty)$ be a finite interval with $\rho \in L_q(I)$. Then for λ defined by (6.25) we have*

(6.27) $$\limsup_{n \to \infty} n^\lambda a_n(T_\rho) \leq 2\, C_{p,q} \|\rho\|_r$$

for $q < \infty$. If $q = \infty$, then (6.27) holds with $\|\rho^*\|_r$ on the right hand side.

PROOF. Given $\varepsilon > 0$, by Proposition 6.4 we find a partition $\mathcal{I} = \{I_1, \ldots, I_N\}$ of I such that

(6.28) $$\left(\sum_{k=1}^N J_\rho(I_k)^r\right)^{1/r} \leq (1 + \varepsilon) \|\rho\|_r .$$

Now let

$$T_\rho = T_\rho P^\circ + T_\rho P^c$$

with P°, P^c projecting $L_p(I)$ onto $L_p^\circ(I; \mathcal{I})$ and $L_p^c(I; \mathcal{I})$, respectively. Note that $\mathrm{rank}(T_\rho P^c) \leq N$ implies

$$a_n(T_\rho P^c) = 0 \quad \text{for} \quad n \geq N .$$

Consequently, using the additivity of the approximation numbers, by Corollary 6.7 and by (6.28) we get

$$\limsup_{n \to \infty} n^\lambda a_n(T_\rho) = \limsup_{n \to \infty} n^\lambda a_n(T_\rho P^\circ)$$

$$\leq 2 \cdot C_{p,q} \left(\sum_{k=1}^N J_\rho(I_k)^r\right)^{1/r} \leq 2 \cdot C_{p,q} (1 + \varepsilon) \|\rho\|_r .$$

Letting $\varepsilon \to 0$ immediately proves the theorem. □

6.3. Estimates for $a_n(T_\rho)$ in the Case $1 < p < 2 < q < \infty$

If we compare the results in Theorem 6.8 with those in Theorem 2.1, we see that the power λ in Theorem 6.8 cannot be optimal for $1 \leq p < 2 < q \leq \infty$. To get the optimal order of $a_n(T_\rho)$ in this case as well, more involved approximation constructions are necessary. We have to use here quite sophisticated tools as multi-level partitions and Gluskin's bound for embeddings of finite-dimensional l_p–spaces. Unfortunately, these methods do not apply to the cases $p = 1$, $2 < q < \infty$ and $1 < p < 2$, $q = \infty$, thus the exact behaviour of $a_n(T_\rho)$ remains open for those p and q.

Our basic result in the case $1 < p < 2 < q < \infty$ is as follows.

PROPOSITION 6.9. *Let I be a finite interval, $\rho \in L_q(I)$ and regard T_ρ as operator from $L_p(I)$ to $L_q(I)$ with $1 < p < 2 < q < \infty$. Then there exist an integer $n_0 = n_0(p,q)$ and a constant $c = c(p,q)$, both written explicitly below, such that for each $n \geq n_0$*

$$(6.29) \qquad a_{3n}(T_\rho) \leq 2\, C_{p,q} J_\rho(I)\, (1 + c^{(G)} c)\, n^{-\lambda}.$$

Here $C_{p,q}$ is Maz'ja-Rosin constant from (6.2), $c^{(G)}$ is Gluskin's constant from Lemma 6.10 below, and λ is defined by

$$(6.30) \qquad \lambda := \min\{1/p', 1/q\} + 1/2\,.$$

Remark: Notice that in the given range of parameters
$$(6.31) \qquad \begin{aligned} \lambda &> \min\{1/p', 1/q\} + \max\{1/p', 1/q\} = 1/p' + 1/q \\ &= 1/r = \min\{1, 1/r\}\,, \end{aligned}$$

consequently estimate (6.29) is stronger than the general one of (6.27).

PROOF. *Step 1. Partitions.* For given n we construct a family of increasing partitions $\mathcal{I}_0, \mathcal{I}_1, \ldots$ of the interval I. We will construct

$$\mathcal{I}_d := \{I_{d,j}\}, \quad 0 \leq j < 2^d n, \quad d \geq 0,$$

as follows. Let $\mathcal{I}_0$ be a partition of I in n pieces such that (see (6.12) and (6.8))

$$J_\rho(I_{0,1}) = \cdots = J_\rho(I_{0,n}) \leq n^{-1/r} J_\rho(I).$$

Furthermore, once $\mathcal{I}_d$ is constructed, we split each interval $I_{d,j}$ into two intervals,

$$I_{d,j} = I_{d+1,2j} \cup I_{d+1,2j+1}\,,$$

so that $J_\rho(I_{d+1,2j}) = J_\rho(I_{d+1,2j+1})$. Using (6.8) and an induction argument, we have

$$(6.32) \qquad \max_j J_\rho(I_{d+1,j}) \leq 2^{-1/r} \max_j J_\rho(I_{d,j}) \leq \cdots \leq 2^{-(d+1)/r} n^{-1/r} J_\rho(I).$$

Step 2. Factor-spaces and factorizing operators. Now define for each $d \geq 0$ finite dimensional spaces by

$$R_d := \Big\{ f \in L_p(I): f \text{ is } \mathcal{I}_{d+1}\text{-measurable and } \int_{I_{d,j}} f(t)\,dt = 0\,,\ 0 \leq j < 2^d n \Big\}.$$

An equivalent definition is

$$R_d := \operatorname{span}\{\varphi_{d,j},\ 0 \leq j < 2^d n\}$$

where

$$\varphi_{d,j} = b_{d,j} \left(\frac{\mathbf{1}_{I_{d+1,2j}}}{|I_{d+1,2j}|} - \frac{\mathbf{1}_{I_{d+1,2j+1}}}{|I_{d+1,2j+1}|} \right)$$

and the constants $b_{d,j}$ are chosen to provide $\|\varphi_{d,j}\|_p = 1$. Notice that each $\varphi_{d,j}$ is supported by its own interval $I_{d,j}$. In the same way we set

$$Q_d := T_\rho(R_d) = \operatorname{span}\{T_\rho(\varphi_{d,j}),\ 0 \leq j < 2^d n\}$$

and by $T_\rho^{(d)} : R_d \to Q_d$ we denote the restriction of T_ρ to R_d.
Next we introduce the operator

$$\pi_d^R : R_d \to l_p^{2^d n} \quad \text{with} \quad \pi_d^R \left(\sum_{j < 2^d n} x_j \varphi_{d,j} \right) = (x_j)\,,$$

the identity
$$E_d : l_p^{2^d n} \to l_q^{2^d n},$$
and, finally, we define $\pi_d^Q : l_q^{2^d n} \to Q_d$ by
$$\pi_d^Q(y) = \sum_{j<2^d n} y_j T_\rho \varphi_{d,j}, \quad y = (y_j) \in l_q^{2^d n}.$$

Of course, the key relation between these operators is
$$T_\rho^{(d)} = \pi_d^Q E_d \, \pi_d^R.$$

Notice that π_d^R is an isometry and that for each $y \in l_q^{2^d n}$ we have

(6.33)
$$\left\|\pi_d^Q(y)\right\|_q^q = \sum_{j<2^d n} |y_j|^q \, \|T_\rho \varphi_{d,j}\|_q^q.$$

By Corollary 6.2 it follows that
$$\|T_\rho \varphi_{d,j}\|_q \leq C_{p,q} \, J_\rho(I_{d,j}),$$
which yields, by virtue of (6.32) and (6.33),
$$\left\|\pi_d^Q(y)\right\|_q \leq C_{p,q} \, 2^{-d/r} n^{-1/r} J_\rho(I) \, \|y\|_q$$
for each $y \in l_q^{2^d n}$ or, equivalently,
$$\left\|\pi_d^Q\right\| \leq C_{p,q} 2^{-d/r} n^{-1/r} J_\rho(I).$$

Hence, if $\nu \in \mathbb{N}$, then

$$a_\nu(T_\rho^{(d)}) \leq \left\|\pi_d^R\right\| a_\nu(E_d) \left\|\pi_d^Q\right\|$$
(6.34)
$$\leq C_{p,q} \, 2^{-d/r} \, n^{-1/r} J_\rho(I) \cdot a_\nu(E_d)$$

suggests that bounds for the approximation numbers of the identity E_d may be crucial later on. The following lemma is a (slightly simplified) part of a deep result in this direction due to E. D. Gluskin ([**23**]). He calls the approximation numbers "linear widths" and denotes them by d_n' in analogy with the Kolmogorov widths d_n.

LEMMA 6.10. *Let $1 \leq p \leq 2 \leq q \leq \infty$ and let $id_m : l_p^m \to l_q^m$ be the identity operator. Then there exists a constant $c^{(G)} = c^{(G)}(p,q)$ such that for all positive integers n and m the following are valid.*

(1) *For $q \leq p'$*
(6.35)
$$a_n(id_m) \leq c^{(G)} \left(m^{1/q - 1/p} + m^{1/q} n^{-1/2} \right).$$

(2) *If $q \geq p'$, then*
(6.36)
$$a_n(id_m) \leq c^{(G)} \left(m^{1/p' - 1/q'} + m^{1/p'} n^{-1/2} \right).$$

Step 3. Operator decompositions. Now we construct a decomposition of T_ρ via the operators $T_\rho^{(d)}$. To do so let $\mathcal{A}_d : L_p(I) \to L_p(I)$ denote the averaging operators,
$$\mathcal{A}_d f = \sum_{j<2^d n} \frac{\int_{I_{d,j}} f(t) dt}{|I_{d,j}|} \mathbf{1}_{I_{d,j}}, \quad d \geq 0.$$

Define also the increments $\mathcal{C}_d = \mathcal{A}_{d+1} - \mathcal{A}_d : L_p(I) \to R_d$. It is well known that $\|\mathcal{A}_d\| = 1$, hence $\|\mathcal{C}_d\| \leq 2$. Our next calculations depend on an integer parameter $M = M(n)$ specified later on. With this M we write a decomposition of the identity operator $E : L_p(I) \to L_p(I)$ as

$$E = \mathcal{A}_0 + \sum_{d=0}^{M-1} \mathcal{C}_d + (E - \mathcal{A}_M)$$

implying

(6.37) $$T_\rho = T_\rho \mathcal{A}_0 + \sum_{d=0}^{M-1} T_\rho \mathcal{C}_d + T_\rho(E - \mathcal{A}_M) \,.$$

Notice that

(6.38) $$\operatorname{rank}(T_\rho \mathcal{A}_0) \leq \operatorname{rank}(\mathcal{A}_0) = n$$

and that the operator $E - \mathcal{A}_M$ maps into the space $L_p^\circ(I; \mathcal{I}_M)$. Then Lemma 6.5 and (6.32) let us conclude

(6.39)
$$\|T_\rho(E - \mathcal{A}_M)\| \leq \|E - \mathcal{A}_M\| \; C_{p,q} \max_{j < 2^M n} J_\rho(I_{M,j}) \leq C_{p,q} \, 2^{1-M/r} n^{-1/r} J_\rho(I) \,.$$

Now we may write down the bounds for the approximation numbers. For all ν and d it follows that

(6.40) $$a_\nu(T_\rho \mathcal{C}_d) = a_\nu(T_\rho^{(d)} \mathcal{C}_d) \leq a_\nu(T_\rho^{(d)}) \|\mathcal{C}_d\| \leq 2 a_\nu(T_\rho^{(d)}).$$

Next, we define integers ν_d, $0 \leq d < M$, by $\nu_d = [n(d^2+1)^{-1}]$. Then we have

(6.41) $$\sum_{d=0}^{M-1} \nu_d < n \sum_{d=0}^\infty (d^2+1)^{-1} \leq 2\,n.$$

It follows from (6.37) – (6.41) that

(6.42) $$a_{3n}(T_\rho) \leq 2 \sum_{d=0}^{M-1} a_{\nu_d}(T_\rho^{(d)}) + C_{p,q} 2^{1-M/r} n^{-1/r} J_\rho(I) \,.$$

Let us now specify M in order to control the last term in (6.42). Notice first that by (6.31)

$$\alpha := r\lambda - 1 > 0\,.$$

Hence we may choose a positive integer M so that $2^M \in [n^\alpha, 2n^\alpha]$ yielding

$$(2^M n)^{-1/r} \leq n^{-(\alpha+1)/r} = n^{-\lambda} \,.$$

Observe that the last term in (6.42) is bounded by $2\, C_{p,q}\, J_\rho(I)\, n^{-\lambda}$. Next let n_0 be so big that for all $n \geq n_0$

$$\left(\frac{\alpha \log n}{\log 2}\right)^2 + 1 \leq n.$$

For $n \geq n_0$ we infer from $2n^\alpha \geq 2^M$ that $(M-1)^2 + 1 \leq n$, hence we have $\nu_d \geq 1$ for all $d < M$. Now we are in position to apply Lemma 6.10. Assume $q \leq p'$. Using (6.35) with $n = \nu_d$ and $m = 2^d n$ it follows that

$$a_{\nu_d}(E_d) \leq c^{(G)} \left((2^d n)^{1/q - 1/p} + (2^d n)^{1/q} \nu_d^{-1/2} \right),$$

and by (6.34) this implies

$$\sum_{d=0}^{M-1} a_{\nu_d}(T_\rho^{(d)})$$

$$\leq C_{p,q} \, c^{(G)} \, J_\rho(I) \sum_{d=0}^{\infty} \left((2^d n)^{1/q-1/p-1/r} + (2^d n)^{1/q-1/r} \cdot (2(d^2+1)/n)^{1/2} \right).$$

Notice that $1/q - 1/p - 1/r = -1$, so the first part of the last sum may be estimated (recall $\lambda < 1$) by

$$\sum_{d=0}^{\infty} (2^d n)^{1/q-1/p-1/r} = \sum_{d=0}^{\infty} (2^d n)^{-1} = n^{-1} \leq n^{-\lambda} \, .$$

For the second part of the sum, observe that $1/q - 1/r - 1/2 = -1/p' - 1/2 = -\lambda$ and we obtain the bound

$$n^{-\lambda} \sum_{d=0}^{\infty} 2^{-d/p'} (2(d^2+1))^{1/2} = n^{-\lambda} \sum_{d=0}^{\infty} 2^{-d \min\{1/p', 1/q\}} (2(d^2+1))^{1/2} := \tilde{c} \, n^{-\lambda} \, ,$$

where it is important that $1/p' > 0$, i.e. $p > 1$, to conclude $\tilde{c} < \infty$. [2] Summing up our estimates we obtain the asserted bound for $a_{3n}(T_\rho)$ with $c := \tilde{c} + 1$.

Similarly, using (6.36) instead of (6.35), with the same choice of the parameters ν_d we get the claimed estimate for $q \geq p'$. □

Remark: Enlarging the constant in (6.29), by the monotinicity of the a_n's it follows that

(6.43) $$a_n(T_\rho) \leq c \, n^{-\lambda} \, J_\rho(I)$$

for **all** $n \in \mathbb{N}$ and with $c > 0$ only depending on p and q. Recall that Corollary 6.2 implies $a_1(T_\rho) = \|T_\rho\| \leq C_{p,q} J_\rho(I)$.

Having finished the interval–based bounds, we can state and prove the comprehensive analogues of Theorem 6.6, Corollary 6.7 and Theorem 6.8. Hence let us start with extending Theorem 6.6 for p's and q's with $1 < p < 2 < q < \infty$.

THEOREM 6.11. *Suppose $1 < p < 2 < q < \infty$ and define λ by (6.31). Let $I \subseteq (0, \infty)$ be a finite or infinite interval and let as before $\mathcal{I} = \{I_k\}_{k \in \mathcal{K}}$ be a partition of I. If $n_k \in \mathbb{N}$, $k \in \mathcal{K}$, satisfy*

(6.44) $$n := \sum_{k \in \mathcal{K}} (n_k - 1) + 1 < \infty \, ,$$

then

$$a_n(T_\rho^\circ) \leq c \cdot \sup_{k \in \mathcal{K}} n_k^{-\lambda} \, J_\rho(I_k) \, .$$

Here $T_\rho^\circ : L_p^\circ(I; \mathcal{I}) \to L_q(I)$ denotes the restriction of T_ρ to $L_p^\circ(I; \mathcal{I})$.

PROOF. Set $\mathcal{K}_0 := \{k \in \mathcal{K} : n_k > 1\}$, which is by (6.44) necessarily finite. According to (6.43) for each $k \in \mathcal{K}_0$ we find operators $S_k : L_p(I_k) \to L_q(I_k)$ such that $\operatorname{rank}(S_k) \leq n_k - 1$ and, if $f \in L_p(I_k)$, then

$$\|T_\rho f - S_k f\|_{L_q(I_k)} \leq 2 \, a_{n_k}(T_\rho : L_p(I_k) \to L_q(I_k)) \, \|f\|_{L_p(I_k)}$$
$$\leq c \, n_k^{-\lambda} \, J_\rho(I_k) \cdot \|f\|_{L_p(I_k)} \, .$$

---

[2] This is the crucial point where our method does not apply in the case $p = 1$.

Given $f \in L_p^\circ(I;\mathcal{I})$, we define now $Sf \in L_q(I)$ by
$$Sf := \sum_{k \in \mathcal{K}_0} S_k(f \mathbf{1}_{I_k}),$$
and for those f's we obtain

$$\|T_\rho f - Sf\|_q \leq \left(\sum_{k \in \mathcal{K}_0} \|T_\rho(f \mathbf{1}_{I_k}) - S_k(f \mathbf{1}_{I_k})\|_q^q\right)^{1/q} + \left(\sum_{k \notin \mathcal{K}_0} \|T_\rho(f \mathbf{1}_{I_k})\|_q^q\right)^{1/q}$$
$$\leq c \cdot \sup_{k \in \mathcal{K}_0} n_k^{-\lambda} J_\rho(I_k) \cdot \|f\|_p + c \cdot \sup_{k \notin \mathcal{K}_0} J_\rho(I_k) \cdot \|f\|_p$$
$$\leq c \cdot \sup_{k \in \mathcal{K}} n_k^{-\lambda} J_\rho(I_k) \cdot \|f\|_p$$

using $p \leq q$ as well as Corollary 6.2. Since
$$\operatorname{rank}(S) \leq \sum_{k \in \mathcal{K}_0} (n_k - 1) = n - 1,$$
this proves the theorem. $\square$

COROLLARY 6.12. *For $1 < p < 2 < q < \infty$ let I and $\mathcal{I} = \{I_k\}_{k \in \mathcal{K}}$ be as before. Then for the restriction $T_\rho^\circ : L_p^\circ(I;\mathcal{I}) \to L_q(I)$ it follows that*
$$a_n(T_\rho^\circ) \leq c \cdot n^{-\lambda} \left(\sum_{k \in \mathcal{K}} J_\rho(I_k)^{1/\lambda}\right)^\lambda$$
with $\lambda < 1$ defined by (6.30).

PROOF. The proof follows by the same arguments as used in the verification of Corollary 6.7. The only difference is that here we choose n_k's satisfying
$$n_k - 1 < n \cdot \frac{J_\rho(I_k)^{1/\lambda}}{\sum_{j \in \mathcal{K}} J_\rho(I_j)^{1/\lambda}} \leq n, \quad k \in \mathcal{K}.$$
$\square$

To formulate the next result we need the following quantity of $\rho \in L_q(I)$. Here $I \subset (0, \infty)$ is a finite interval. For λ given by (6.30) we set
$$\alpha_\lambda(\rho) := \inf\left\{\left(\sum_{k=1}^N J_\rho(I_k)^{1/\lambda}\right)^\lambda : \mathcal{I} = \{I_1, \ldots, I_N\} \text{ partition of } I\right\}.$$

Remark: Note that $1/\lambda < r$, hence in view of (6.11) it follows that
$$\|\rho\|_r \leq \alpha_\lambda(\rho) \leq J_\rho(I).$$
It would be interesting to find an explicit formula for calculating $\alpha_\lambda(\rho)$.

Now we state and prove the announced extension of Theorem 6.8.

THEOREM 6.13. *Let $I \subset (0, \infty)$ be a finite interval and let $\rho \in L_q(I)$. If $1 < p < 2 < q < \infty$, then with λ given by (6.30) we have*
$$\limsup_{n \to \infty} n^\lambda a_n(T_\rho) \leq c \cdot \alpha_\lambda(\rho).$$

Since the proof is exactly as that of Theorem 6.8, we omit it.

6.4. Approximation Numbers of T_ρ (General Case)

The following result in [18] was the starting point of our investigations.

THEOREM 6.14. *Suppose $T_\rho : L_p(0,\infty) \to L_p(0,\infty)$ with $1 < p < \infty$ and $\rho \in L_p(0,\infty)$. If $|\rho|_1 < \infty$, then*

$$c_1 \cdot \|\rho\|_1 \le \liminf_{n\to\infty} n\, a_n(T_\rho) \le \limsup_{n\to\infty} n\, a_n(T_\rho) \le c_2 \cdot \|\rho\|_1 \;.$$

Remark: Note that for $p = q$ necessarily $r = \lambda = 1$.

The aim of this section is to extend this result to operators T_ρ mapping $L_p(0,\infty)$ into $L_q(0,\infty)$. Moreover, we do not assume $\rho \in L_q(0,\infty)$, only $\rho \in L_q(x,\infty)$ for $x > 0$. Let us start with some notation. Recall that the Δ_k's are the dyadic intervals $[2^k, 2^{k+1})$. If $0 < v < \infty$, then we set

$$|\rho|_v := \left(\sum_{k\in\mathbb{Z}} J_\rho(\Delta_k)^v\right)^{1/v},$$

which for $v = r$ with $1/r = 1/p' + 1/q$ coincides with the former definition of $|\rho|_r$, cf. (2.9). Furthermore, let us set as before

(6.45) $\quad \lambda := \begin{cases} 1 & : 1 \le q < p \le \infty, \\ 1/r & : 1 \le p \le q \le 2,\; 2 \le p \le q \le \infty, \\ 1/2 + \min\{1/p', 1/q\} & : 1 \le p \le 2 \le q \le \infty. \end{cases}$

With this notation the following holds.

THEOREM 6.15. *For $T_\rho : L_p(0,\infty) \to L_q(0,\infty)$ and λ defined by (6.45) the following are valid:*

(1) *If $1 \le q < p \le \infty$, $1 \le p \le q \le 2$ or $2 \le p \le q \le \infty$, then*

(6.46) $$\sup_n n^\lambda a_n(T_\rho : L_p(0,\infty) \to L_q(0,\infty)) \le c\, |\rho|_r \;.$$

(2) *In the case $1 < p < 2 < q < \infty$ we have*

(6.47) $$\sup_n n^\lambda a_n(T_\rho : L_p(0,\infty) \to L_q(0,\infty)) \le c\, |\rho|_{1/\lambda} \;.$$

Remark: Note that in the second case $1/\lambda < r$, hence $|\rho|_r \le |\rho|_{1/\lambda}$, so that the dependence on ρ in (6.47) is weaker than in (6.46).

PROOF. We use similar ideas to those in section 4.2. Let $\Delta := \{\Delta_k\}_{k\in\mathbb{Z}}$ be the partition of $(0,\infty)$ generated by dyadic intervals and let P° and P^c be the projections from $L_p(0,\infty)$ onto $L_p^\circ(\mathbb{R}^+; \Delta)$ and $L_p^c(\mathbb{R}^+; \Delta)$. As before we split T_ρ into two pieces as

$$T_\rho = T_\rho^\circ P^\circ + T^c P^c\;,$$

with the restrictions T° and T^c on the corresponding subspaces. For these two operators the following are true.

PROPOSITION 6.16. *Regard T_ρ° and T_ρ^c as operators mapping from $L_p^\circ(\mathbb{R}^+; \Delta)$ or $L_p^c(\mathbb{R}^+; \Delta)$ into $L_q(0,\infty)$. Then with λ given by (6.45) the following holds:*

(1) *If $1 < p < 2 < q < \infty$, then*

(6.48) $$\sup_n n^\lambda a_n(T_\rho^\circ) \leq c \, |\rho|_{1/\lambda} \; ,$$

while in the remaining cases (we exclude $p = 1$, $2 < q \leq \infty$ and $q = \infty$, $1 < p < 2$)

(6.49) $$\sup_n n^\lambda a_n(T_\rho^\circ) \leq c \, |\rho|_r \; .$$

(2) *For all p, q with $1/r = 1/p' + 1/q > 0$ we have*

(6.50) $$\sup_n n^\lambda a_n(T_\rho^c) \leq c \, |\rho|_{r,\infty} \; .$$

Remark: Since $|\rho|_r \leq |\rho|_{1/\lambda}$, in view of

$$\begin{aligned} a_{n+m-1}(T_\rho) &\leq a_n(T_\rho^\circ P^\circ) + a_m(T_\rho^c P^c) \\ &\leq 2\, a_n(T_\rho^\circ) + a_m(T_\rho^c) \end{aligned}$$

the preceding proposition completely proves Theorem 6.15.

PROOF OF (6.48) AND (6.49): This is an immediate consequence of Corollaries 6.7 and 6.12 when applied to $I = (0, \infty)$ and $\mathcal{I} = \Delta$. □

PROOF OF (6.50): We use here similar ideas to those in the proof of (4.14) in Proposition 4.3. Our notation will be as in this proof. Let $\varphi_p : l_p(\mathbb{Z}) \to L_p^c(0, \infty)$ be the isometry in (4.15) and define the operators R_1 and R_2 from l_p into $L_q(0, \infty)$ as in (4.17) and (4.18), respectively. Then we have (cf. (4.16))

$$T_\rho^c \circ \varphi_p = R_1 + R_2 \; ,$$

and it suffices to estimate the approximation numbers of R_1 and R_2. Let $\mathcal{D} : l_p(\mathbb{Z}) \to l_q(\mathbb{Z})$ be the diagonal operator with diagonal elements $\delta_k = J_\rho(\Delta_k), k \in \mathbb{Z}$. As shown in 4.2, there exist bounded operators j_q and Y from $l_q(\mathbb{Z})$ into $L_q(0, \infty)$ defined in (4.23) and (4.19) as well as an operator $V : l_p(\mathbb{Z}) \to l_p(\mathbb{Z})$ given in (4.21) such that

$$R_1 = Y \circ \mathcal{D} \quad \text{and} \quad R_2 = j_q \circ \mathcal{D} \circ V \; .$$

Consequently, it remains to estimate $a_n(\mathcal{D} : l_p \to l_q)$ suitably. Here we use a general result about approximation numbers of diagonal operators (cf. [**36**]).

Let $(\sigma_i)_{i \geq 1}$ be a sequence of real numbers and let $S : l_p \to l_q$ be the associated diagonal operator. Let $r, u, v \in [1, \infty]$ and suppose that

(6.51) $$1/r > (1/q - 1/p)_+ \; .$$

Then $(\sigma_i) \in l_{r,v}$ implies $(a_n(S)) \in l_{u,v}$ in each of the following cases:

(A) $1/u = 1/r + 1/p - 1/q$ for $1 \leq q \leq p \leq \infty$.
(B) $1/u = 1/r$ for $1 \leq p \leq q \leq 2$ or $2 \leq p \leq q \leq \infty$.
(C) $1/u = 1/r + 1/2 - \max\{1/p', 1/q\}$ if $1 - 1/r \leq 1/p$, $1/q \leq 1/r$ and $1 \leq p \leq 2 \leq q \leq \infty$.

Moreover, if these conditions are satisfied, then there are universal constants $c > 0$ such that

$$\|(a_n(S))\|_{u,v} \leq c \cdot \|(\sigma_i)\|_{r,v} \; .$$

In the special case $1/r = 1/p' + 1/q$, condition (6.51) is satisfied (cf. proof of (4.14)) and so are the conditions for p and q in (C). Thus in this situation we get the following:

If $1/r = 1/p' + 1/q$ and $(\sigma_i) \in l_{r,\infty}$, then this implies $(a_n(S)) \in l_{u,\infty}$ with u's as follows:

(A) $u = 1$ for $1 \leq q \leq p \leq \infty$.
(B) $u = r$ for $1 \leq p \leq q \leq 2$ or $2 \leq p \leq q \leq \infty$.
(C) $1/u = 1/2 + \min\{1/p', 1/q\}$ for $1 \leq p \leq 2 \leq q \leq \infty$.

Now we apply this result (for r as before) to the diagonal operators $\mathcal{D}^+$ and $\mathcal{D}^-$ generated by the sequences $(J_\rho(\Delta_k))_{k \geq 0}$ and $(J_\rho(\Delta_k))_{k < 0}$ and with λ defined by (6.45). Hence

$$\sup_n n^\lambda a_n(\mathcal{D}^+) \leq c \cdot \|(J_\rho(\Delta_k))_{k \geq 0}\|_{r,\infty}$$

as well as a similar estimate for $a_n(\mathcal{D}^-)$. From these one easily derives (6.50), completing the proof of Proposition 6.16 and, hence, also of Theorem 6.15. □

Now we may state and prove the announced generalization of Theorem 6.14.

THEOREM 6.17. *Regard T_ρ as operator from $L_p(0, \infty)$ into $L_q(0, \infty)$ and define λ by (6.45).*

(1) *If either $1 \leq q < p \leq \infty$, $1 \leq p \leq q \leq 2$ or $2 \leq p \leq q \leq \infty$, then $|\rho|_r < \infty$ implies*

(6.52)
$$\limsup_{n \to \infty} n^\lambda a_n(T_\rho) \leq 2\,C_{p,q}\,\|\rho\|_r$$

with $\|\rho^*\|_r$ on the right hand side of (6.52) for $q = \infty$.

(2) *If $1 < p < 2 < q < \infty$ and $|\rho|_{1/\lambda} < \infty$, then*

(6.53)
$$\limsup_{n \to \infty} n^\lambda a_n(T_\rho) \leq c \cdot \inf_{\mathcal{I}} \left(\sum_{k \in \mathcal{K}} J_\rho(I_k)^{1/\lambda} \right)^\lambda$$

where the inf in (6.53) is taken over all countable partitions $\mathcal{I} = \{I_k\}_{k \in \mathcal{K}}$ of $(0, \infty)$ with

$$\operatorname{card}\{k \in \mathcal{K} : I_k \cap A \neq \emptyset\} < \infty$$

for each $A \subset (0, \infty)$ compact.

PROOF. The proof of the first case follows exactly as that for Theorem 2.2 (2) by applying Theorems 6.8 and 6.15. Recall that one splits ρ as $\rho = \rho_1 + \rho_2$ with $\rho_1 \in L_q$ supported on some finite interval and $|\rho_2|_r < \varepsilon$ for given $\varepsilon > 0$.

To verify (6.53) a similar technique is used. For $\varepsilon > 0$ we choose a finite interval $I \subset (0, \infty)$ such that for $\rho_1 := \rho \cdot \mathbf{1}_I$ and $\rho_2 := \rho \cdot \mathbf{1}_{I^c}$ we have $\rho_1 \in L_q(I)$ and $|\rho_2|_{1/\lambda} < \varepsilon$. Especially, in view of Theorem 6.15 we get

$$\sup_n n^\lambda a_n(T_{\rho_2}) \leq c\varepsilon\,.$$

Thus using the additivity of the a_n's, it remains to estimate $\limsup_n n^\lambda a_n(T_{\rho_1})$ suitably. To do so let $\mathcal{I} = \{I_k\}_{k \in \mathcal{K}}$ be an arbitrary partition of $(0, \infty)$ possessing the properties as in (6.53). By assumption

$$I \cap \mathcal{I} := \{I \cap I_k\}_{k \in \mathcal{K}}$$

defines a finite partition of I, thus Theorem 6.13 applies and leads to

$$\limsup_{n\to\infty} n^\lambda a_n(T_{\rho_1}) \leq c \cdot \left(\sum_{k\in\mathcal{K}} J_{\rho_1}(I_k \cap I)^{1/\lambda}\right)^\lambda$$

$$\leq c \cdot \left(\sum_{k\in\mathcal{K}} J_\rho(I_k)^{1/\lambda}\right)^\lambda.$$

Taking in the last estimate the inf over all partitions $\mathcal{I}$ completes the proof of (6.53). $\square$

Remarks:
(1) It remains open whether in the second case $|\rho|_r < \infty$ suffices for the validity of (6.53) (maybe even with $\|\rho\|_r$ on the right hand side). Note that $|\rho|_{1/\lambda} < \infty$ implies $|\rho|_r < \infty$ in the given range of p's and q's.
(2) In view of estimate (2.1), Theorem 2.3 yields the following: If $1 \leq q \leq p \leq \infty$ and $p > 1$, then there are functions $\rho \in L_r(0,\infty)$ such that

$$\limsup_{n\to\infty} n\, a_n(T_\rho) = \infty$$

(note that in this case $\lambda = 1$). Especially, this applies for $p = q$ showing that Theorem 9 in [18] fails for arbitrary ρ's in $L_1(0,\infty)$. It would be interesting to know whether or not $|\rho|_r < \infty$ (or possibly $|\rho|_{1/\lambda} < \infty$) is also necessary in the remaining cases of p's and q's.

COROLLARY 6.18. *Let $1 \leq q < p \leq \infty$, $1 \leq p \leq q \leq 2$ or $2 \leq p \leq q \leq \infty$ and suppose that there are some $\alpha, \beta \in \mathbb{R}$ such that $t^\alpha \rho(t)$ is monotone near zero and $t^\beta \rho(t)$ is so near infinity. Then it follows that*

$$\limsup_{n\to\infty} n^\lambda a_n(T_\rho) \leq 2\, C_{p,q} \, \|\rho\|_r$$

with the usual modification for $q = \infty$. Moreover, the assumption about monotonicity near zero may be replaced by $\rho \in L_q(0,\infty)$.

PROOF. This is a direct consequence of Corollary 5.12 and Theorem 6.17. $\square$

Next we state the corresponding estimates for the lower limit. Since the exact behaviour of $a_n(T_1 : L_p(0,1) \to L_q(0,1))$ seems to be unknown for $p = 1$, $2 < q < \infty$ and $q = \infty$, $1 < p < 2$, (cf. Theorem 2.1) we have to exclude those combinations of p's and q's here.

THEOREM 6.19. *Let $1 \leq p, q \leq \infty$ with $1 \leq q \leq 2$ for $p = 1$ and $2 \leq p \leq \infty$ if $q = \infty$. Then it follows that*

$$(6.54) \quad c \cdot \|\rho\|_r \leq \liminf_{n\to\infty} n^\lambda a_n(T_\rho : L_p(0,\infty) \to L_q(0,\infty))$$

with λ defined by (6.45). If $q = \infty$, the left hand norm in (6.54) may be replaced by $\|\rho^*\|_{p'}$. The constant $c > 0$ in (6.54) may be chosen as

$$(6.55) \quad c = (1/2) \liminf_{n\to\infty} n^\lambda a_n(T_1 : L_p(0,1) \to L_q(0,1))$$

(for $p = 2$ the additional factor $1/2$ is not necessary). Especially, under the assumptions of Corollary 6.18 it follows that

$$(6.56) \quad c \cdot \|\rho\|_r \leq \liminf_{n\to\infty} n^\lambda a_n(T_\rho) \leq \limsup_{n\to\infty} n^\lambda a_n(T_\rho) \leq 2\, C_{p,q} \, \|\rho\|_r$$

with $\|\rho^*\|_{p'}$ on both sides for $q = \infty$.

PROOF. This follows from Proposition 5.4 exactly as the corresponding assertion (Theorem 2.4, (2)) for the entropy numbers. Note that we used only properties of the entropy numbers valid also for the a_n's. For $q = \infty$, replace T_ρ by T_{ρ^*} and recall (cf. (2.11) and its consequences) that $a_n(T_\rho) = a_n(T_{\rho^*})$ in this case. □

Remark: For $1 < p = q < \infty$ in [**18**] a lower bound of the constant $c > 0$ in (6.54) was given. For example, if $p = q = 2$, then c may be chosen as $1/\pi$ while our upper estimate gives $C_{2,2} = 1$ (recall that the factor 2 on the right hand side of (6.56) is not necessary for $p = 2$) which is weaker than the corresponding estimate in [**18**].

Our final objective is to transform estimates for $a_n(T_\rho)$ into those of $a_n(T_{\rho,\psi})$. Before doing so, let us shortly recall some basic facts about the underlying scale transformation θ defined by

$$\theta(s) = \int_0^s \psi(t)^{p'} \, dt, \quad s > 0,$$

with range $A = (0, \theta(\infty))$ and pseudo–inverse $\theta^- : A \to (0, \infty)$ as in (3.23). If $\tilde{\rho}$ on A is given by (3.24), i.e.

$$\tilde{\rho}(\tau) = \rho(\theta^-\tau)\, \psi(\theta^-\tau)^{-p'/q}, \quad \tau \in A,$$

equality (3.9) yields the following:
If $\tilde{I} \subseteq A$ is some finite interval with end points $\theta(a)$ and $\theta(b)$, then we have

(6.57) $$J_{\tilde{\rho}}(\tilde{I}) = \|\psi\|_{L_{p'}(a,b)} \, \|\rho\|_{L_q(a,b)}.$$

Furthermore let us recall that $\delta_k(\rho, \psi)$ was defined in (3.35) and that with $1/r = 1/p' + 1/q$ we set

$$|(\rho, \psi)|_r = \left(\sum_{k \in \mathbb{Z}} \delta_k(\rho, \psi)^r \right)^{1/r}.$$

Using (3.39), (6.57) and the methods in the proof of Theorem 4.6, the preceding estimates for $a_n(T_\rho)$ lead to the following bounds for $a_n(T_{\rho,\psi})$ in the case $p > 1$. Recall that the above transformation applies for $p > 1$ only.

THEOREM 6.20. *Let ρ, ψ be as before and suppose $p > 1$. If $1 \leq q \leq p \leq \infty$, $1 < p \leq q \leq 2$ or $2 \leq p \leq q \leq \infty$, then with $\lambda = \min\{1, 1/r\}$ the following are valid:*

(1) *For $T_{\rho,\psi} : L_p(0, \infty) \to L_q(0, \infty)$ it follows that*

$$\sup_n n^\lambda \, a_n(T_{\rho,\psi}) \leq c \cdot |(\rho, \psi)|_r.$$

(2) *If $I \subseteq (0, \infty)$ is a bounded interval and $\rho \in L_q(I)$, $\psi \in L_{p'}(I)$, then we have*

(6.58) $$\limsup_{n \to \infty} n^\lambda \, a_n(T_{\rho,\psi}) \leq 2\, C_{p,q} \cdot \|\rho\,\psi\|_r$$

with $1/r = 1/p' + 1/q$.

(3) *If $|(\rho, \psi)|_r < \infty$ and $T_{\rho,\psi} : L_p(0, \infty) \to L_q(0, \infty)$, then it also follows that*

(6.59) $$\limsup_{n \to \infty} n^\lambda \, a_n(T_{\rho,\psi}) \leq 2\, C_{p,q} \cdot \|\rho\,\psi\|_r$$

with $\|\rho^*\,\psi\|_{p'}$ in (6.58) and (6.59) for $q = \infty$.

Suppose now $1 < p < 2 < q < \infty$ and define λ as before by $\lambda = 1/2 + \min\{1/p', 1/q\}$.

(4) For $T_{\rho,\psi}: L_p(0,\infty) \to L_q(0,\infty)$ it follows that

$$\sup_n n^\lambda a_n(T_{\rho,\psi}) \leq c \cdot \left(\sum_{k \in \mathbb{Z}} \delta_k(\rho,\psi)^{1/\lambda} \right)^\lambda. \tag{6.60}$$

(5) If $I \subseteq (0,\infty)$ is a bounded interval and $\rho \in L_q(I)$, $\psi \in L_{p'}(I)$, then we have

$$\limsup_{n \to \infty} n^\lambda a_n(T_{\rho,\psi}) \leq c \cdot \inf_{\mathcal{I}} \left\{ \left(\sum_{k=1}^N \|\psi\|_{L_{p'}(I_k)}^{1/\lambda} \cdot \|\rho\|_{L_q(I_k)}^{1/\lambda} \right)^\lambda \right\}$$

where the inf is taken over all finite partitions $\mathcal{I} = \{I_1, \ldots, I_N\}$ of I.

(6) Suppose that the right hand expression of (6.60) is finite and regard $T_{\rho,\psi}$ as operator from $L_p(0,\infty)$ into $L_q(0,\infty)$. Then it follows that

$$\limsup_{n \to \infty} n^\lambda a_n(T_{\rho,\psi}) \leq c \cdot \inf_{\mathcal{I}} \left\{ \left(\sum_{k \in \mathcal{K}} \|\psi\|_{L_{p'}(I_k)}^{1/\lambda} \cdot \|\rho\|_{L_q(I_k)}^{1/\lambda} \right)^\lambda \right\}$$

where the inf is taken over partitions $\mathcal{I}$ of $(0,\infty)$ as in (2) of Theorem 6.17.

The corresponding lower estimates are derived from Theorem 6.19 and (3.39).

THEOREM 6.21. For $p > 1$, $1 \leq q \leq \infty$ and $q < \infty$ for $1 < p < 2$ it follows that

$$c \, \|\rho\psi\|_r \leq \liminf_{n \to \infty} n^\lambda a_n(T_{\rho,\psi})$$

with c given by (6.55) and λ by (6.45). Again we may replace the left hand side by $\|\rho^*\psi\|_{p'}$ for $q = \infty$.

Under additional assumptions about ρ and ψ the following two–sided estimates are valid.

COROLLARY 6.22. Under the regularity conditions (1) or (2) of Corollary 5.13 we have the following. If $p > 1$ and $1 \leq q \leq p \leq \infty$, $1 < p \leq q \leq 2$ or $2 \leq p \leq q \leq \infty$, then with $\lambda = \min\{1, 1/r\}$ and the usual modification for $q = \infty$ we have

$$c \, \|\rho\psi\|_r \leq \liminf_{n \to \infty} n^\lambda a_n(T_{\rho,\psi}) \leq \limsup_{n \to \infty} n^\lambda a_n(T_{\rho,\psi}) \leq 2\, C_{p,q} \cdot \|\rho\psi\|_r.$$

Finally, let us treat the case $p = 1$ not covered by the preceding results. As for entropy numbers we use a duality argument which applies here more directly because of $a_n(T) = a_n(T')$ for compact operators T (cf. [**38**], 11.7.4). Recall that we needed for the e_n's the quite sophisticated Proposition 4.11 to apply this argument.

THEOREM 6.23. Suppose $1 \leq q \leq 2$ and let $T_{\rho,\psi}: L_1(0,\infty) \to L_q(0,\infty)$.

(1) Then

$$\sup_n n^{1/q} a_n(T_{\rho,\psi}) \leq c \cdot \left(\sum_{k \in \mathbb{Z}} 2^k \|\psi\|_{L_\infty(v_{k+1}, v_k)}^q \right)^{1/q} \tag{6.61}$$

where the v_k's are maximal numbers satisfying

$$\int_{v_k}^\infty \rho(t)^q \, dt = 2^k, \quad k \in \mathbb{Z}.$$

(2) *Whenever the right hand side of (6.61) is finite, then*

(6.62) $$\limsup_{n\to\infty} n^{1/q} a_n(T_{\rho,\psi}) \leq 2 \cdot \|\rho\psi^*\|_q$$

with ψ^* defined in (2.11).

(3) *Furthermore, we always have*

$$c \cdot \|\rho\psi^*\|_q \leq \liminf_{n\to\infty} n^{1/q} a_n(T_{\rho,\psi})$$

with $c > 0$ defined in (6.55) for $p = 1$.

(4) *If ψ is monotone near zero and infinity, then*

(6.63) $$c \cdot \|\rho\psi^*\|_q \leq \liminf_{n\to\infty} n^{1/q} a_n(T_{\rho,\psi}) \leq \limsup_{n\to\infty} n^{1/q} a_n(T_{\rho,\psi}) \leq 2 \cdot \|\rho\psi^*\|_q$$

PROOF. We use similar arguments as in the proof of Theorem 4.9. Since, as mentioned above, $a_n(T) = a_n(T')$ for compact T's, it suffices to verify all assertions for the dual $S_{\psi,\rho} : L_{q'} \to L_\infty$ of $T_{\rho,\psi}$. However $S_{\psi,\rho}$ is isomorphic to $T_{\tilde\psi,\tilde\rho}$ defined in Proposition 3.7 and mapping $L_{q'}(0,\infty)$ to $L_\infty(0,\infty)$. Because of $2 \leq q' \leq \infty$, Theorems 6.20 and 6.21 as well as Corollary 6.22 apply to $T_{\tilde\psi,\tilde\rho}$ and complete the proof. Let us still note that $C_{q',\infty} = 1$ (cf. Remark after Theorem 6.1), so that only the constant 2 appears on the right hand sides of (6.62) and (6.63), respectively. □

CHAPTER 7

Small Ball Behaviour of Weighted Wiener Processes

This chapter is devoted to applications of our preceding results in the theory of Gaussian stochastic processes. Hence we start with introducing some basic notions and results about those processes. For further information we refer to [34], [22], [27] or [28].

7.1. Gaussian Processes and Metric Entropy

Let E be a (real) Banach space and let X be an E–valued random variable, i.e. X is measurable with respect to the Borel σ–algebra $\mathcal{B}(E)$ on E and almost surely X attains its values in a separable subspace of E. As usual $\mathbb{P}_X$ denotes the distribution of X, i.e. $\mathbb{P}_X$ is the measure on $(E, \mathcal{B}(E))$ generated by X.

The random variable X is said to be Gaussian centered provided that for each $x' \in E'$, the dual of E, the real valued random variable $x'(X)$ is centered normal on $\mathbb{R}$ (we do not exclude the case $x'(X) = 0$ a.s.). Suppose now we are given a separable Hilbert space H and a bounded operator T from H into E. Let $(f_k)_{k=1}^\infty$ be an ONB in H and let $\xi_1, \xi_2, \ldots$ be an i.i.d. sequence of real valued standard normal random variables. Whenever

$$(7.1) \qquad \sum_{k=1}^\infty \xi_k T f_k$$

converges a.s. in E, then this sum defines a centered Gaussian E–valued random variable X. Conversely, given a centered Gaussian random variable X, there exist a Hilbert space H and an operator T from H into E such that the sum (7.1) exists a.s. in E and possesses the same distribution as X. It is worthwhile to mention that the distribution of the random sum (7.1) is independent of the special choice of the ONB and, moreover, whenever (7.1) exists a.s. for one ONB, then this is valid for every ONB in H.

Each E–valued Gaussian random variable X generates a subset $K_X \subset E$ (unit ball of the reproducing kernel Hilbert space of X) by

$$K_X := \left\{ z \in E : |x'(z)| \leq \left(\mathbb{E} |x'(X)|^2 \right)^{1/2} ; \; x' \in E' \right\}.$$

The set K_X is known to be compact in E (cf. [22]). For X represented by an operator T via (7.1), it follows that

$$\begin{aligned} K_X &= \{ z \in E : |x'(z)| \leq \|T'x'\| ; \; x' \in E' \} \\ &= \{ Tf : f \in H; \; \|f\|_H \leq 1 \} \end{aligned}$$

which, for example, implies

$$e_n(T) = \inf\left\{\varepsilon > 0 : K_X \subseteq \bigcup_{j=1}^{2^{n-1}} B(z_j;\varepsilon);\ z_1,\ldots,z_{2^{n-1}} \in K_X\right\}.$$

In other words, the entropy numbers of T describe the degree of compactness of the set K_X generated by the random variable X. There exist tight and important relations between the metric entropy of K_X (or, equivalently of T) and properties of X. Combining results of R. M. Dudley [14] and V. N. Sudakov [42] with those of [44] the following is valid: The sum (7.1) exists a.s. provided that $\sum_n n^{-1/2} e_n(T) < \infty$ while, conversely, the a.s. convergence of (7.1) necessarily yields $\sup_n n^{1/2} e_n(T) < \infty$. A faster decay of the entropy numbers of T is reflected by a "better" small ball behaviour of X as follows (cf. [26] and [32]).

THEOREM 7.1. *Let T be an operator from H into E such that the Gaussian random variable X is well-defined via (7.1). For $0 < \alpha < 2$ the following statements are equivalent:*

(i) *We have*

(7.2) $$d_1 := \limsup_{n\to\infty} n^{1/\alpha}\, e_n(T) < \infty\,.$$

(ii) *We have*

(7.3) $$\liminf_{\varepsilon\to 0} \varepsilon^{\frac{2\alpha}{2-\alpha}} \log \mathbb{P}(\|X\| \leq \varepsilon) := -\kappa_1 > -\infty\,.$$

Conversely, if T satisfies

(7.4) $$d_2 := \liminf_{n\to\infty} n^{1/\alpha}\, e_n(T) > 0\,,$$

then this implies

(7.5) $$\limsup_{\varepsilon\to 0} \varepsilon^{\frac{2\alpha}{2-\alpha}} \log \mathbb{P}(\|X\| \leq \varepsilon) := -\kappa_2 < 0\,.$$

Remark: As recently proved (cf. [12]), there are universal positive constants c_1 and c_2 depending only on α such that for d_1 in (7.2) and κ_1 in (7.3)

$$\kappa_1 \leq c_1\, d_1^{\frac{2\alpha}{2-\alpha}}$$

and if (7.4) holds, then κ_2 in (7.5) satisfies

$$\kappa_2 \geq c_2\, d_2^{\frac{2\alpha}{2-\alpha}}\,.$$

For later application let us also mention a special case of the relation between linear operators on Hilbert spaces and Gaussian random variables. Suppose that T is a linear map from H into $C[0,\infty)$, the space of continuous functions on $[0,\infty)$. If for some ONB $(f_k)_{k=1}^\infty$ in H the series (7.1) converges uniformly on $[0,a]$ for all $a > 0$, then all restrictions of X to $[0,a]$ are $C[0,a]$-valued random variables. Moreover, we can consider X as a random variable with values in the locally convex space $C[0,\infty)$. Setting

$$X(s) := \sum_{k=1}^\infty \xi_k (Tf_k)(s), \quad s \geq 0,$$

we may also regard X as centered Gaussian stochastic process, indexed by $[0,\infty)$ and possessing a.s. continuous sample paths. In this case, the distribution of X is uniquely determined by its covariance function

$$(t,s) \to \mathbb{E}\, X(t)X(s), \quad t,s \geq 0\,.$$

7.2. Weighted Wiener Processes

Theorem 7.1 suggests that Theorems 2.2 and 2.4 as well as their generalizations to $T_{\rho,\psi}$ imply certain small ball estimates for $L_q(0,\infty)$–valued random variables generated by T_ρ or $T_{\rho,\psi}$, respectively. In order to apply the former entropy estimates to small ball problems, the operators have to be defined on a Hilbert space, i.e. we have to regard here T_ρ and $T_{\rho,\psi}$ as operators from $L_2(0,\infty)$ into $L_q(0,\infty)$ with $1 \leq q \leq \infty$. Then the number $r > 0$ equals

$$1/r = 1/2 + 1/q, \quad \text{i.e.} \quad r = \frac{2q}{2+q}\,.$$

Given an ONB $(f_k)_{k=1}^\infty$ in $L_2(0,\infty)$ let us assume that

(7.6) $$W_{\rho,\psi} := \sum_{k=1}^\infty \xi_k\, T_{\rho,\psi} f_k$$

converges a.s. in $L_q(0,\infty)$. Then $W_{\rho,\psi}$ defines an $L_q(0,\infty)$–valued centered Gaussian random variable. To describe it, let us first consider the ordinary integration operator T_1. Regarding T_1 as mapping from $L_2(0,\infty)$ into $C[0,\infty)$, it is well–known that for each ONB $(f_k)_{k=1}^\infty$ the sum

$$\sum_{k=1}^\infty \xi_k\, T_1 f_k$$

converges a.s. uniformly on compact subsets of $[0,\infty)$, hence it generates a centered Gaussian process W with a.s. continuous sample paths via

$$W(s) = \sum_{k=1}^\infty \xi_k (T_1 f_k)(s) = \sum_{k=1}^\infty \xi_k \int_0^s f_k(t)\, dt, \quad s \geq 0\,.$$

Direct calculations give

$$\mathbb{E}\, W(t)W(s) = \min\{t,s\} \quad \text{for} \quad t,s \geq 0\,,$$

i.e. W is the classical Wiener process over $[0,\infty)$ with distribution $\mathbb{P}_W$, the classical Wiener measure on $C[0,\infty)$.

We give now a similar interpretation for operators $T_{1,\psi}$. Let $\psi \in L_2(0,x)$ for each $x > 0$. Then we define the increasing scale transformation θ as in Example 3.3(a) with $p=2$ by

$$\theta(s) = \int_0^s \psi(t)^2\, dt\,.$$

With $N_\theta = \{t > 0 : \psi(t) > 0\}$ and $E_2^\theta \subseteq L_2(0,\infty)$ defined by (3.12), the mapping $T_{1,\psi}$ may be regarded on E_2^θ with values in $C[0,\infty)$. Recall that $T_{1,\psi}$ vanishes on the orthogonal complement of E_2^θ. Given an ONB $(f_k)_{k=1}^\infty$ in E_2^θ, by Proposition 3.1 the functions $g_k := (\Phi_2^\theta)^{-1} f_k$ form an ONB in $L_2(0,\theta(\infty))$. Using the definitions (3.10), (3.24), (3.19) and Proposition 3.2, we have for every k

$$(T_{1,\psi} f_k)(s) = \Phi_q^\theta T_{\tilde{\mathbf{1}},\tilde{\psi}}\, g_k(s) = (T_1 g_k)(\theta(s))\,.$$

Hence,

$$W_{1,\psi}(s) := \sum_{k=1}^{\infty} \xi_k (T_{1,\psi} f_k)(s)$$

$$= \sum_{k=1}^{\infty} \xi_k (T_1 g_k)(\theta(s)),$$

and it follows that

$$W_{1,\psi}(s) = W(\theta(s)), \quad s \geq 0,$$

for a Wiener process W on $[0, \theta(\infty))$. In other words, Gaussian variables allowing a series representation (7.1) with respect to some $T_{1,\psi}$ are those exactly of the form $X(s) = W(\theta(s))$, $s \geq 0$, with θ increasing, absolutely continuous and $\theta' \in L_1(0, x)$ for each $x > 0$.

Given a non–negative function ρ with $\rho \in L_q(x, \infty)$ for each $x > 0$, the series

(7.7) $$\sum_{k=1}^{\infty} \xi_k (T_\rho f_k) = \sum_{k=1}^{\infty} \xi_k \rho (T_1 f_k)$$

converges a.s. pointwise to ρW. It converges a.s. in $L_q(0, \infty)$ for some ONB $(f_k)_{k=1}^{\infty}$ in $L_2(0, \infty)$ iff

$$\mathbb{P}_W(w \in C[0, \infty) : \rho w \in L_q) = 1 .$$

In other words, (7.7) converges a.s. in L_q iff

$$\mathbb{P}(\|\rho W\|_q < \infty) = 1$$

and it follows that

$$W_{\rho,1}(s) = \rho(s) W(s)$$

for a Wiener process W on $[0, \infty)$. Combining both cases, the sum (7.6) converges a.s. in L_q iff

$$\mathbb{P}(\|\rho W(\theta(\,\cdot\,))\|_q < \infty) = 1$$

and

(7.8) $$W_{\rho,\psi}(s) = \rho(s) W(\theta(s))$$

for $s \geq 0$.

Remark: It is known that for $1 \leq q < \infty$ the sum (7.6) converges a.s. in $L_q(0, \infty)$ iff

$$\int_0^\infty \rho(s)^q \theta(s)^{q/2} \, ds = \int_0^{\theta(\infty)} \tilde{\rho}(\tau)^q \tau^{q/2} \, d\tau < \infty .$$

For $q = \infty$ a similar integral description is known as the Kolmogorov–Petrovskii–Erdös–Feller test. It holds only under some additional monotonicity assumptions about $\tilde{\rho}$ (cf. [**24**]). For example, if $\theta(t) = t$, then we have $\rho W \in L_\infty$ iff for some $R > 0$

$$\int_0^\infty \exp\left\{-\frac{R}{\rho(s)^2 s}\right\} \frac{ds}{s} < \infty.$$

7.3. Small Ball Estimates for Wiener Processes

Let us start with some classical estimates for the small ball behaviour of the Wiener process over $[0,1]$. We refer to [35] or [33] for more detailed information.

PROPOSITION 7.2. *We have*
$$\mathbb{P}(\sup_{0\leq s\leq 1} |W(s)| \leq \varepsilon) \sim \frac{4}{\pi} \exp(-\frac{\pi^2}{8}\varepsilon^{-2})$$

and, if $1 \leq q < \infty$, there are positive constants $c(q)$ and k_q such that
$$\mathbb{P}\Big(\int_0^1 |W(s)|^q \, ds \leq \varepsilon^q\Big) \sim c(q)\,\varepsilon \, \exp(-k_q\, \varepsilon^{-2}) \,.$$

For $q = 2$ we have $c(2) = 4/\pi^{1/2}$ and $k_2 = 1/8$.

Remark: For exact formula of the supremum distribution, see [11] or [21], v.2, Chapter X. The small ball behavior of the L^2-norm for a Wiener process was obtained by Cameron and Martin [6] in 1944, while Borovkov and Mogulskii proved an asymptotic formula for general L_q–norms much later, see [4].

The results stated in Proposition 7.2 imply, in particular,

(7.9) $$\log \mathbb{P}(\|W\|_{L_q(0,1)} \leq \varepsilon) \sim -k_q\, \varepsilon^{-2}$$

with $k_q > 0$ as above ($k_\infty = \pi^2/8$). Let us still mention the following for later application. If
$$B(s) := W(s) - W(1), \quad 0 \leq s \leq 1,$$
denotes a Brownian bridge over $[0,1]$, then

(7.10) $$\log \mathbb{P}(\|B\|_{L_q(0,1)} \leq \varepsilon) \sim -k_q\, \varepsilon^{-2}$$

as well with same constants $k_q > 0$.

It is natural to ask for a characterization of functions $\rho \in L_q(x, \infty)$, $x > 0$, such that the weighted process $(\rho(s)\, W(s))_{0 \leq s < \infty}$ possesses a small ball behaviour similar to that of $(W(s))_{0 \leq s \leq 1}$. More precisely, one may ask which ρ's satisfy with some $\kappa_1 > 0$

(7.11) $$\log \mathbb{P}(\|\rho \cdot W\|_{L_q(0,\infty)} \leq \varepsilon) \geq -\kappa_1\, \varepsilon^{-2}$$

as $\varepsilon < \varepsilon_0$ or for ρ's such that with some $\kappa_2 > 0$

(7.12) $$\log \mathbb{P}(\|\rho \cdot W\|_{L_q(0,\infty)} \leq \varepsilon) \leq -\kappa_2\, \varepsilon^{-2}$$

as $\varepsilon < \varepsilon_0$. In view of Theorem 7.1 we see that (7.11) is equivalent to
$$\limsup_{n\to\infty} n\, e_n\big(T_\rho : L_2(0,\infty) \to L_q(0,\infty)\big) < \infty \,,$$
while (7.12) holds provided that
$$\liminf_{n\to\infty} n\, e_n\big(T_\rho : L_2(0,\infty) \to L_q(0,\infty)\big) > 0 \,.$$

Hence the former entropy estimates imply the following results concerning (7.11).

THEOREM 7.3. *Let $\rho \geq 0$ be as before and for $1 \leq q \leq \infty$ let $r > 0$ be defined by $1/r = 1/2 + 1/q$.*

(1) If $|\rho|_r < \infty$, then for $1 \leq q < \infty$ and for some $c = c(q) > 0$

$$\log \mathbb{P}(\|\rho W\|_{L_q(0,\infty)} \leq \varepsilon) \geq -c \cdot \varepsilon^{-2} \|\rho\|_r^2$$

while

$$\log \mathbb{P}(\sup_{0 \leq t < \infty} |\rho(t) W(t)| \leq \varepsilon) \geq -c \cdot \varepsilon^{-2} \|\rho^*\|_2^2$$

as $\varepsilon < \varepsilon_0$.

(2) For any $q \in [1,\infty]$ there are functions $\rho \in L_r(0,\infty) \cap L_q(0,\infty)$ such that (7.11) does not hold, i.e. we have

(7.13) $$\liminf_{\varepsilon \to 0} \varepsilon^2 \log \mathbb{P}(\|\rho W\|_q \leq \varepsilon) = -\infty.$$

PROOF. Assertions (1) and (2) follow directly via Theorem 7.1 by Theorem 2.2 (3) or Theorem 2.3 (1), respectively. □

Remark: A careful inspection of the proof of Theorem 2.3 (1) shows that the function ρ constructed there lives on $(1,\infty)$ only. Hence, using the transformation presented in example 3.3 (b), we get also ρ's in L_r, failing (7.11) and with $\rho(t) = 0$ for $t \in [1,\infty)$. For example, if $q = \infty$, then there exists a $\rho \in L_2(0,1)$, bounded on $(x,1]$ for every $x > 0$, such that

$$\liminf_{\varepsilon \to 0} \varepsilon^2 \log \mathbb{P}(\sup_{0 \leq s \leq 1} |\rho(s) W(s)| \leq \varepsilon) = -\infty.$$

The following conditions are necessary for the validity of (7.11).

THEOREM 7.4. *Suppose that $\rho \geq 0$ satisfies (7.11) with some $\kappa_1 > 0$, i.e. we have*

$$\log \mathbb{P}(\|\rho W\|_{L_q(0,\infty)} \leq \varepsilon) \geq -\kappa_1 \cdot \varepsilon^{-2}$$

as $\varepsilon < \varepsilon_0$ for some $\kappa_1 > 0$. Then necessarily

$$\rho \in L_r(0,\infty) \quad \text{with} \quad \|\rho\|_r^2 \leq c(q) \cdot \kappa_1$$

and, furthermore,

$$|\rho|_{r,\infty} < \infty,$$

where as before $1/r = 1/2 + 1/q$.

PROOF. The results follow immediately from Theorem 2.4 via Theorem 7.1. □

The two preceding theorems show that we do not have a complete characterization of functions ρ satisfying (7.11). There is a small gap between necessary and sufficient conditions expressed by $|\rho|_{r,\infty} < \infty$ and $|\rho|_r < \infty$, respectively. Yet using Corollary 2.5, under some additional monotonicity assumptions the following is true.

COROLLARY 7.5. *Suppose $\rho \in L_q(0,\infty)$ or that for some $\alpha \in \mathbb{R}$ the function $t^\alpha \rho(t)$ is monotone near zero and that, furthermore, for some $\beta \in \mathbb{R}$ the function $t^\beta \rho(t)$ is monotone near infinity. Then the following statements are equivalent.*

(i) There is a $\kappa_1 > 0$ with

$$\log \mathbb{P}(\|\rho W\|_q \leq \varepsilon) \geq -\kappa_1 \varepsilon^{-2}$$

for $\varepsilon > 0$ small.

(ii) The function ρ satisfies
$$\int_0^\infty \rho(t)^{\frac{2q}{2+q}}\,dt < \infty\,.$$

Next we investigate the validity of (7.12). Here Theorem 7.1 leads to the following consequence of Theorem 2.4.

THEOREM 7.6. *Let as before $1/r = 1/2 + 1/q$. Then for any $\rho \geq 0$ we have*
$$\log \mathbb{P}(\|\rho W\|_q \leq \varepsilon) \leq -c(q) \cdot \|\rho\|_r^2 \cdot \varepsilon^{-2}\,,$$
$\varepsilon < \varepsilon_0$ *Consequently, if $|\rho|_r < \infty$ (or $\rho \in L_r$ satisfies the assumptions of Corollary 7.5), then*
$$\log \mathbb{P}(\|\rho W\|_q \leq \varepsilon) \approx -\|\rho\|_r^2 \cdot \varepsilon^{-2}$$
as $\varepsilon \to 0$.

All preceding results have natural extensions to more general processes defined by (7.8) with θ increasing, absolutely continuous and $\theta' \in L_1(0,x)$ for each $x > 0$ (θ' corresponds to ψ^2). For example, the following holds by virtue of Theorems 4.6 and 7.1.

THEOREM 7.7. *Let θ be increasing, absolutely continuous with $\theta' \in L_1(0,x)$ for $x > 0$ and let $\rho \in L_q(x,\infty)$ for $x > 0$. Let $u_k > 0$ be the minimal solutions of*

(7.14) $$\theta(u_k) = 2^k,\quad k \in \mathbb{Z},$$

with $u_k = \infty$ if there is no solution of (7.14). If

(7.15) $$\sum_{k=-\infty}^{\infty} 2^{\frac{kq}{2+q}} \|\rho\|_{L_q(u_k,u_{k+1})}^{\frac{2q}{2+q}} < \infty\,,$$

then

(7.16) $$\log \mathbb{P}(\|\rho W(\theta(\cdot))\|_q \leq \varepsilon) \approx -\|\rho\sqrt{\theta'}\|_r^2 \cdot \varepsilon^{-2}$$

as $\varepsilon \to 0$ (with ρ^* on the right hand side of (7.16) for $q = \infty$).

The following special case may be of interest.

COROLLARY 7.8. *Suppose that θ' and ρ in Theorem 7.7 are disjointly supported and satisfy condition (7.15). Then this implies*
$$\lim_{\varepsilon \to 0} \varepsilon^2 \log \mathbb{P}(\|\rho W(\theta(\cdot))\|_q \leq \varepsilon) = 0\,.$$

7.4. Exact Small Ball Estimates

As shown in Theorem 7.6, the former entropy estimates imply

(7.17) $$-c_1 \cdot \|\rho\|_r^2 \leq \liminf_{\varepsilon \to 0} \varepsilon^2 \log \mathbb{P}(\|\rho W\|_q < \varepsilon)$$
$$\leq \limsup_{\varepsilon \to 0} \varepsilon^2 \log \mathbb{P}(\|\rho W\|_q < \varepsilon) \leq -c_2 \cdot \|\rho\|_r^2$$

provided that $|\rho|_r < \infty$. Recall that $1/r = 1/2 + 1/q$. Here $0 < c_2 \leq c_1 < \infty$ are some universal constants depending only on q. Further information about the best possible values of c_1 and c_2 cannot be derived from entropy estimates. We want to show now how this can be done by probabilistic methods. More precisely, we will prove $c_1 = c_2 = k_q$ with k_q given in (7.9). The previous work was, however, not useless, since (7.17) will play an important role to suppress some remainder terms coming from the tails of the kernel functions.

THEOREM 7.9. *Let $1 \leq q \leq \infty$ and define $r < \infty$ as before by $1/r = 1/2 + 1/q$. If $|\rho|_r < \infty$, then*

$$\lim_{\varepsilon \to 0} \varepsilon^2 \log \mathbb{P}(\|\rho W\|_q \leq \varepsilon) = -k_q \cdot \|\rho\|_r^2, \quad q < \infty, \quad \text{and}$$

$$\lim_{\varepsilon \to 0} \varepsilon^2 \log \mathbb{P}(\|\rho W\|_\infty \leq \varepsilon) = -(\pi^2/8) \cdot \|\rho^*\|_2^2.$$

PROOF. The proof is divided into three steps. In the first one we show the following.

PROPOSITION 7.10. *For $1 \leq q \leq \infty$ and $\rho \geq 0$ as before we have*
(7.18) $$\limsup_{\varepsilon \to 0} \varepsilon^2 \log \mathbb{P}(\|\rho W\|_q \leq \varepsilon) \leq -k_q \cdot \|\rho\|_r^2.$$

PROOF. Of course, it suffices to verify (7.18) for $\rho \in L_q(I)$ with some finite interval $I \subset (0, \infty)$. The proof of (7.18) will follow the ideas and methods developed in section 5, but this time from a probabilistic point of view[1]. Hence we start with an application of Lemma 5.1 which asserts the following. Given $\delta, \sigma > 0$ we find an i.s.f. φ of the form

$$\varphi = \sum_{j=1}^{N} \alpha_j \mathbf{1}_{I_j}, \quad \alpha_j \geq 0,$$

such that for
(7.19) $$G := \{s \in I : \rho(s) > (1-\delta)\varphi(s)\}$$
we have
(7.20) $$\int_G \varphi(s)^r \, ds \geq \int_I \rho(s)^r \, ds - \sigma.$$

Assume now $I_j = [a_j, b_j]$ and define stochastic processes B_j over I_j by

$$B_j(t) := W(t) - W(a_j) - \frac{t - a_j}{b_j - a_j}(W(b_j) - W(a_j))$$

for $t \in I_j$. (We call such a process a Brownian bridge over I_j).

It is known (cf. [**34**], Section 3), that $B_1, \ldots, B_N$ are independent and, moreover, we have
(7.21) $$W(t) = B_j(t) + Y(t), \quad t \in I_j,$$
with some Gaussian process Y independent of $B_1, \ldots, B_N$. Setting $B(t) := B_j(t)$ for $t \in I_j$, Anderson's inequality (cf. [**1**] or [**34**], Section 11) implies then

(7.22) $$\begin{aligned} \mathbb{P}(\|\rho W\|_q \leq \varepsilon) &\leq \mathbb{P}(\|\rho B\|_q \leq \varepsilon) \leq \mathbb{P}(\|\rho \mathbf{1}_G B\|_q \leq \varepsilon) \\ &\leq \mathbb{P}(\|\varphi \mathbf{1}_G B\|_q \leq \varepsilon(1-\delta)^{-1}) \end{aligned}$$

where the last step follows from (7.19). Let φ_G be defined as in (5.6), i.e. we have

$$\varphi_G = \varphi \cdot \mathbf{1}_G = \sum_{j=1}^{N} \alpha_j \cdot \mathbf{1}_{I_j \cap G},$$

so that we may rewrite (7.22) as
(7.23) $$\mathbb{P}(\|\rho W\|_q \leq \varepsilon) \leq \mathbb{P}(\|\varphi_G B\|_q \leq \varepsilon(1-\delta)^{-1}).$$

---

[1] Obviously, the entropy approach via Theorem 7.1 is not precise enough to obtain exact values for the bounds we are interested in.

Next define $L_2^\circ(I)$ by (3.31) with respect to the partition $\{I_1, \ldots, I_N\}$ and let as before $T_{\varphi_G}^\circ$ be the restriction of T_{φ_G} to the Hilbert space $L_2^\circ(I)$. The operator $T_{\varphi_G}^\circ$ generates via (7.1) an $L_q(I)$–valued Gaussian random variable, which, as easily can be seen, has the same distribution as $\varphi_G B$. Regard now a projector $S : L_2^\circ(I) \to L_2^\circ(I)$ with $Sf := \mathbf{1}_G f$. Then by Anderson inequality we obtain

$$(7.24) \qquad \mathbb{P}(\|\varphi_G B\|_q \leq \varepsilon(1-\delta)^{-1}) \leq \mathbb{P}(\|\varphi_G B_G\|_q \leq \varepsilon(1-\delta)^{-1})$$

where $\varphi_G B_G$ denotes the Gaussian process generated via (7.1) by $T_{\varphi_G}^\circ \circ S = T_{\varphi_G,\mathbf{1}_G}^\circ$. Now we may apply Proposition 5.3 and see that $T_{\varphi_G,\mathbf{1}_G}^\circ$ is isomorphic to $T_{\tilde\varphi_G}^\circ$ mapping $L_2^\circ(0,|G|)$ into $L_q(0,|G|)$. Recall the definition of $\tilde\varphi_G$. We divide the interval $(0,|G|)$ into N intervals $G_1, \ldots, G_N$ with length $|G_j| = |I_j \cap G|$, $1 \leq j \leq N$. Then

$$\tilde\varphi_G = \sum_{j=1}^N \alpha_j \mathbf{1}_{G_j}.$$

In the language of processes the isomorphism of linear operators means

$$\|\varphi_G B_G\|_q \stackrel{d}{=} \left\|\sum_{j=1}^N \alpha_j \tilde B_j\right\|_q$$

where $\tilde B_1, \ldots, \tilde B_N$ are independent Brownian bridges on $G_1, \ldots, G_N$, respectively. A final scale transformation leads to

$$\|\varphi_G B_G\|_q \stackrel{d}{=} \left\|\sum_{j=1}^N \alpha_j |G_j|^{1/r} \bar B_j\right\|_q$$

with $\bar B_1, \ldots, \bar B_N$ independent standard Brownian bridges on $[0,1]$. Now (7.23) and (7.24) imply

$$(7.25) \qquad \mathbb{P}(\|\rho W\|_q \leq \varepsilon) \leq \mathbb{P}\Big(\sum_{j=1}^N \alpha_j^q |G_j|^{q/r} \eta_j \leq \varepsilon^q (1-\delta)^{-q}\Big)$$

where the i.i.d. random variables η_j are defined by

$$(7.26) \qquad \eta_j := \int_0^1 |\bar B_j(t)|^q \, dt, \quad 1 \leq j \leq N,$$

with obvious modification for $q = \infty$. We know from (7.10) that

$$\lim_{\varepsilon \to 0} \varepsilon^2 \log \mathbb{P}(\eta_j \leq \varepsilon^q) = -k_q,$$

thus Lemma 2.2 in [**30**] implies

$$\lim_{\varepsilon \to 0} \varepsilon^2 \log \mathbb{P}\Big(\sum_{j=1}^N \alpha_j^q |G_j|^{q/r} \eta_j \leq \varepsilon^q (1-\delta)^{-q}\Big)$$

$$= -k_q \cdot \Big(\sum_{j=1}^N \alpha_j^r |G_j|\Big)^{2/r} \cdot (1-\delta)^2$$

$$= -k_q \|\tilde\varphi_G\|_r^2 (1-\delta)^2$$

$$\leq -k_q (\|\rho\|_r^r - \sigma)^{2/r} (1-\delta)^2$$

by (7.20). Combining this with (7.25) finally leads to

$$\limsup_{\varepsilon \to 0} \varepsilon^2 \log \mathbb{P}(\|\rho W\|_q \leq \varepsilon) \leq -k_q \big(\|\rho\|_r^r - \sigma \big)^{2/r} (1-\delta)^2$$

for all $\sigma, \delta > 0$. This completes the proof. □

The next step gives the desired lower estimate for functions ρ defined on finite intervals. More precisely, the following is valid.

PROPOSITION 7.11. *Let $I \subset (0, \infty)$ be a finite interval and let $\rho \in L_q(I)$. Then*

$$\liminf_{\varepsilon \to 0} \varepsilon^2 \log \mathbb{P}(\|\rho W\|_q \leq \varepsilon) \geq -k_q \|\rho\|_r^2 .$$

PROOF. Let us first treat the case $q < \infty$. By Lemma 4.4 we find an i.s.f. φ such that $\|\varphi\|_r = \|\rho\|_r$ and

$$\int_I \big[\rho(s)^q - \varphi(s)^q \big]_+ ds < \sigma$$

for some $\sigma > 0$ specified later on. This implies

$$\|\rho W\|_q^q \leq \int_I \varphi(s)^q |W(s)|^q \, ds + \sup_{s \in I} |W(s)|^q \cdot \int_I \big[\rho(s)^q - \varphi(s)^q \big]_+ ds$$

$$\leq \int_I \varphi(s)^q |W(s)|^q \, ds + \sigma \cdot \sup_{s \in I} |W(s)|^q .$$

Consequently, for all $\varepsilon, \delta > 0$ we have

$$\mathbb{P}(\|\rho W\|_q^q \leq (1+3\delta)\varepsilon^q) \geq \mathbb{P}(A \cap B)$$

where

$$A := \left\{ \int_I \varphi(s)^q |W(s)|^q \, ds \leq (1+2\delta)\varepsilon^q \right\} \quad \text{and}$$

$$B := \left\{ \sup_{s \in I} |W(s)|^q \leq \frac{\delta \varepsilon^q}{\sigma} \right\} .$$

We apply now a partial correlation inequality due to W.V. Li (cf. [31]) and obtain

(7.27) $$\mathbb{P}(A \cap B) \geq \mathbb{P}(\lambda A) \cdot \mathbb{P}\big((1-\lambda^2)^{1/2} B \big)$$

for all $0 < \lambda < 1$ which leads in our case to

$$\mathbb{P}(\|\rho W\|_q^q \leq (1+3\delta)\varepsilon^q) \geq \mathbb{P}\left(\int_I \varphi(s)^q |W(s)|^q \, ds \leq (1+2\delta)\varepsilon^q \lambda^q \right)$$

(7.28) $$\times \mathbb{P}\left(\sup_{0 \leq s \leq 1} |W(s)|^q \leq \frac{\delta \varepsilon^q (1-\lambda^2)^{q/2}}{\sigma |I|^{q/2}} \right) .$$

Next we specify λ and σ in dependence of $\delta > 0$. First we choose λ so close to one that

(7.29) $$(1+2\delta) \lambda^q \geq 1 + \delta$$

and next $\sigma > 0$ with

$$\sigma \leq \delta^{1+q} |I|^{-q/2} (1-\lambda^2)^{q/2}$$

for some λ satisfying (7.29). Then for those λ and σ estimate (7.28) leads to

$$
\begin{aligned}
(7.30) \quad \mathbb{P}(\|\rho W\|_q^q \leq (1+3\delta)\varepsilon^q) &\geq \mathbb{P}\Big(\int_I \varphi(s)^q\,|W(s)|^q\,ds \leq (1+\delta)\varepsilon^q\Big) \\
&\quad \times \mathbb{P}\Big(\sup_{0\leq s\leq 1} |W(s)|^q \leq \delta^{-q}\varepsilon^q\Big).
\end{aligned}
$$

Let us first treat the second term on the right hand side of (7.30). By (7.9)

$$
(7.31) \quad \lim_{\varepsilon\to 0} \varepsilon^2 \log \mathbb{P}\Big(\sup_{0\leq s\leq 1} |W(s)| \leq \delta^{-1}\varepsilon\Big) = -\frac{\pi^2\delta^2}{8},
$$

so it remains to investigate the first term of (7.30). Suppose φ has the representation

$$
\varphi = \sum_{j=1}^N \alpha_j \mathbf{1}_{I_j}, \quad \alpha_j \geq 0.
$$

Let the bridges $B_1,\ldots,B_N$ and Y be as in (7.21). The process φY is generated by $T_\varphi \circ P^c$ (we use the notation as in the proof of Lemma 4.4), i.e. by an operator of rank less or equal than N, hence

$$
(7.32) \quad \mathbb{P}(\|\varphi Y\|_q \leq \varepsilon) \geq c\cdot \varepsilon^N.
$$

Note that

$$
\mathbb{P}\big(\|\varphi W\|_q^q \leq (1+\delta)\varepsilon^q\big) \geq \mathbb{P}\Big(\Big\|\sum_{j=1}^N \alpha_j B_j\Big\|_q \leq \varepsilon\Big) \cdot \mathbb{P}\big(\|\varphi Y\|_q^q \leq \delta\,\varepsilon^q\big),
$$

so (7.32) implies

$$
(7.33) \quad \liminf_{\varepsilon\to 0} \varepsilon^2 \log \mathbb{P}(\|\varphi W\|_q^q \leq (1+\delta)\varepsilon^q) \geq \liminf_{\varepsilon\to 0} \varepsilon^2 \log \mathbb{P}\Big(\Big\|\sum_{j=1}^N \alpha_j B_j\Big\|_q \leq \varepsilon\Big).
$$

Setting

$$
(7.34) \quad \Lambda := \|\rho\|_r^r = \|\varphi\|_r^r = \sum_{j=1}^N \alpha_j^r |I_j|,
$$

with the η_j's given by (7.26) we have

$$
\begin{aligned}
\mathbb{P}\Big(\Big\|\sum_{j=1}^N \alpha_j B_j\Big\|_q \leq \varepsilon\Big) &= \mathbb{P}\Big(\sum_{j=1}^N |I_j|^{q/2+1}\,\alpha_j^q\,\eta_j \leq \varepsilon^q\Big) \\
&\geq \prod_{j=1}^N \mathbb{P}\Big(|I_j|^{q/2+1}\,\alpha_j^q\,\eta_j \leq \frac{\alpha_j^r\,|I_j|}{\Lambda}\cdot \varepsilon^q\Big) \\
&= \prod_{j=1}^N \mathbb{P}\Big(\eta_j \leq \frac{\alpha_j^{r-q}\,|I_j|^{-q/2}}{\Lambda}\cdot \varepsilon^q\Big).
\end{aligned}
$$

Consequently, by (7.10) we get

$$\liminf_{\varepsilon \to 0} \varepsilon^2 \log \mathbb{P}\Big(\Big\|\sum_{j=1}^N \alpha_j B_j\Big\|_q \le \varepsilon\Big) \ge \sum_{j=1}^N \lim_{\varepsilon \to 0} \varepsilon^2 \log \mathbb{P}\Big(\eta_j \le \frac{\alpha_j^{r-q}\, |I_j|^{-q/2}}{\Lambda} \cdot \varepsilon^q\Big)$$

$$= -k_q \cdot \Lambda^{2/q} \cdot \sum_{j=1}^N \alpha_j^{\frac{(q-r)2}{q}} |I_j|$$

$$= -k_q \cdot \Lambda^{2/q+1} = -k_q \cdot \Lambda^{2/r}.$$

From (7.30), (7.31), (7.34) and (7.33) we finally derive

$$\liminf_{\varepsilon \to 0} \varepsilon^2 \log \mathbb{P}(\|\rho W\|_q \le \varepsilon) \ge (1+3\delta)^{-2/q}\Big(-k_q\, \|\rho\|_r^2 - \frac{\pi^2}{8}\delta^2\Big)$$

for every $\delta > 0$. Letting $\delta \to 0$ proves Theorem 7.9 for $q < \infty$.

For $q = \infty$ we apply Lemma 4.5 instead of Lemma 4.4 and obtain an i.s.f. φ with $\rho \le \varphi$ a.e. and

(7.35) $$\int_I \varphi(s)^r \, ds \le \int_I \rho^*(s)^r \, ds + \sigma$$

for $\sigma > 0$ given. This implies

$$\|\rho W\|_\infty \le \|\varphi W\|_\infty,$$

so that we get in this case directly

$$\mathbb{P}(\|\rho W\|_\infty \le \varepsilon) \le \mathbb{P}(\|\varphi W\|_\infty \le \varepsilon).$$

Now the proof can be finished as for $q < \infty$ (without the additional term in (7.30)). A small difference appears in the final estimate where we get only $\|\varphi\|_r^2$, yet by (7.35) this may be replaced by $\|\rho^*\|_r^2$ by letting $\sigma \to 0$. □

Proof of Theorem 7.9: It remains to show that $|\rho|_r < \infty$ implies

(7.36) $$\liminf_{\varepsilon \to 0} \varepsilon^2 \log \mathbb{P}(\|\rho W\|_q \le \varepsilon) \ge -k_q\, \|\rho\|_r^2.$$

Fix a small $\sigma > 0$ and choose disjointly supported functions ρ_1, ρ_2 such that $\rho_1 \in L_q(I)$ for some finite interval $I \subset (0, \infty)$,

(7.37) $$\|\rho_2\|_r \le \sigma \quad \text{and} \quad \rho = \rho_1 + \rho_2.$$

Then for any $\varepsilon > 0$ and $0 < \delta < 1$ it follows that

$$\mathbb{P}(\|\rho W\|_q \le \varepsilon) \ge \mathbb{P}\Big(\|\rho_1 W\|_q \le (1-\delta)\varepsilon,\, \|\rho_2 W\|_q \le \delta \varepsilon\Big)$$

(7.38) $$\ge \mathbb{P}\Big(\|\rho_1 W\|_q \le (1-\delta)\varepsilon \lambda\Big) \cdot \mathbb{P}\Big(\|\rho_2 W\|_q \le \delta \varepsilon (1-\lambda^2)^{1/2}\Big)$$

where we used in the last step again Li's partial correlation inequality (7.27). Recall that λ in (7.38) may be chosen arbitrarily between 0 and 1. Since $|\rho_2|_r \le |\rho|_r < \infty$, the lower bound from (7.17) yields

(7.39) $$\liminf_{\varepsilon \to 0} \varepsilon^2 \log \mathbb{P}\Big(\|\rho_2 W\|_q \le \delta \varepsilon (1-\lambda^2)^{1/2}\Big) \ge -c_1\, \delta^{-2} (1-\lambda^2)^{-1} \cdot \|\rho_2\|_r^2$$

while Proposition 7.11 implies

(7.40) $$\liminf_{\varepsilon \to 0} \varepsilon^2 \log \mathbb{P}\Big(\|\rho_1 W\|_q \le (1-\delta)\varepsilon \lambda\Big) \ge -k_q\, (1-\delta)^{-2} \lambda^{-2} \cdot \|\rho_1\|_r^2$$

$$\ge k_q\, (1-\delta)^{-2} \lambda^{-2} \cdot \|\rho\|_r^2.$$

Finally (7.37), (7.38), (7.39) and (7.40) provide
$$\liminf_{\varepsilon \to 0} \varepsilon^2 \log \mathbb{P}(\|\rho W\|_q \leq \varepsilon) \geq -k_q (1-\delta)^{-2} \lambda^{-2} \cdot \|\rho\|_r^2 - c_1 \delta^{-2} (1-\lambda^2)^{-1} \cdot \sigma^2 .$$
To complete the proof, we first take the limit $\sigma \to 0$, then $\delta \to 0$ and $\lambda \to 1$ and obtain (7.36) as asserted. □

COROLLARY 7.12. *Suppose that $\rho \in L_r(0, \infty)$ satisfies the assumptions of Corollary 2.5. Then*
$$\lim_{\varepsilon \to 0} \varepsilon^2 \log \mathbb{P}(\|\rho W\|_q \leq \varepsilon) = -k_q \cdot \|\rho\|_r^2$$
for $q < \infty$ and
$$\lim_{\varepsilon \to 0} \varepsilon^2 \log \mathbb{P}(\sup_{0<t<\infty} |\rho(t) W(t)| \leq \varepsilon) = -(\pi^2/8) \cdot \|\rho^*\|_2^2 .$$

Remark: Corollary 7.12 was recently proved by W. V. Li in [**29**] and [**30**] (see also [**3**]) for Riemann integrable functions ρ under slightly stronger regularity assumptions at zero and infinity. Notice that we do not need Riemann integrability due to the more flexible procedure of approximation by step functions (Lemmas 4.4, 4.5 and 5.1).

Appendix

The aim of the appendix is to indicate a proof of Theorem 2.1, especially in the critical cases $p \in \{1, \infty\}$ or $q \in \{1, \infty\}$, respectively.

For each pair p, q let $i_{p,q}$ be the natural embedding from the Sobolev space $W_p^1(0,1)$ into $L_q(0,1)$. Then up to an one–dimensional operator (note that the range of T_1 consists of functions in $W_p^1(0,1)$ vanishing at zero) T_1 is isomorphic to $i_{p,q}$, hence for all p, q

$$e_n(T_1 : L_p \to L_q) \approx e_n(i_{p,q}) \,.$$

Using the notation of [**19**], for $1 < p, q < \infty$ we have the identities

(+) $\qquad L_q(0,1) = F_{q,2}^0(0,1) \quad \text{and} \quad W_p^1(0,1) = F_{p,2}^1(0,1) \,,$

thus Theorem 2, p.118 in [**19**] applies and leads to $e_n(i_{p,q}) \approx 1/n$ in this case.

If at least one of the numbers p and q is either 1 or ∞, then (+) is no longer valid, hence Theorem 2 of [**19**] cannot be applied directly. Here we have to use the following inclusions where the corresponding embedding maps are continuous (cf. [**19**], p.44):

$$B_{1,1}^0(0,1) \subset \quad L_1(0,1) \quad \subset B_{1,\infty}^0(0,1) \quad \text{and}$$
$$B_{\infty,1}^0(0,1) \subset \quad L_\infty(0,1) \quad \subset B_{\infty,\infty}^0(0,1) \,,$$

from which one easily derives

$$B_{1,1}^1(0,1) \subset \quad W_1^1(0,1) \quad \subset B_{1,\infty}^1(0,1) \quad \text{and}$$
$$B_{\infty,1}^1(0,1) \subset \quad W_\infty^1(0,1) \quad \subset B_{\infty,\infty}^1(0,1) \,.$$

Note that $W_\infty^1(0,1)$ coincides with the space $C^1(0,1)$ of continuously differentiable functions while $B_{\infty,\infty}^1(0,1)$ consists exactly of 1–Hölder continuous functions (cf. [**19**], p.26). We illustrate as an example how these inclusions can be used to prove (2.4) also for extreme p's and/or q's. Hence let us regard, for example, $p = 1$ and $1 < q < \infty$. Using Theorem 2, p.118 in [**19**] we derive

$$\frac{c_1}{n} \leq e_n(id : B_{1,1}^1 \to L_q) \leq \left\| id : B_{1,1}^1 \to W_1^1 \right\| \cdot e_n(i_{1,q})$$

as well as

$$e_n(i_{1,q}) \leq \left\| id : W_1^1 \to B_{1,\infty}^1 \right\| \cdot e_n(id : B_{1,\infty} \to L_q) \leq \frac{c_2}{n}$$

which, of course, proves (2.4) for $p = 1$ and $1 < q < \infty$. The remaining cases may be proved by similar arguments.

Assertion (2.5) for approximation numbers follows in the same way (by using similar arguments in the critical cases) from Theorem 3.3.4, page 119, in [**19**]. But note, that this theorem does not apply in the cases $p = 1$, $2 < q < \infty$ and $q = \infty$, $1 < p < 2$.

Bibliography

[1] ANDERSON, T. W. (1955). The integral of symmetric unimodular functions over a symmetric convex set and some probability inequalities. *Proc. Amer. Math. Soc.* **6**, 170–176.

[2] BELINSKY, E. S. (1998). Estimates of entropy numbers and Gaussian measures for classes of functions with bounded mixed derivative. *J. Approx. Theor.* **93**, 114-127.

[3] BERTHET, P. and SHI, Z. (2000). Small ball estimates for Brownian motion under a weighted sup-norm. *Studia Sci. Math. Hungarica* **36**, 275-289.

[4] BOROVKOV, A.A. and MOGULSKII A.A. (1991). On probabilities of small deviations. *Siberian Adv. in Math.* **1**, 39–63.

[5] BOURGAIN, J., PAJOR, A., SZAREK, S. and TOMCZAK–JAEGERMANN, N. (1989). On the duality problem for entropy numbers of operators. *Geometric Aspects of Functional Analysis, Lecture Notes in Math.* **1376**, 50-63.

[6] CAMERON, R.H. and MARTIN, W.T. (1944). The Wiener measure of Hilbert neighborhoods in the space of real continuous functions. *J. Math. Phys.* **23**, 195–209.

[7] CARL, B. (1981). Entropy numbers, s–numbers and eigenvalue problems. *J. Funct. Anal.* **41**, 290–306.

[8] CARL, B. (1981). Entropy numbers of diagonal operators with application to eigenvalue problems. *J. Approx. Theor.* **32**, 135–150.

[9] CARL, B., KYREZI, I. and PAJOR, A. (1999). Metric entropy of convex hulls in Banach spaces. *J. London Math. Soc.* **60**, 871–896.

[10] CARL, B. and STEPHANI, I. (1990). Entropy, Compactness and Approximation of Operators. *Cambridge Univ. Press.* Cambridge.

[11] CHUNG, K.L. (1948). On the maximal partial sums of independent random variables. *Trans. Amer. Math. Soc.* **64**, 205–233.

[12] CREUTZIG, J. (1999). Gaußmaße kleiner Kugeln und metrische Entropie. Diplomarbeit, FSU Jena.

[13] CREUTZIG, J. and LINDE, W. (2001). Entropy numbers of certain summation operators. *Georgian Math. Journal* **18**, to appear.

[14] DUDLEY, R. M. (1967). The sizes of compact subsets of Hilbert space and continuity of Gaussian processes. *J. Funct. Anal.* **1**, 290-330.

[15] DUNKER, T. (2000). Estimates for the small ball probabilities of the fractional Brownian sheet. *J. Theor. Probab.* **13**, 357–382.

[16] DUNKER, T., KÜHN, T., LIFSHITS, M. A. and LINDE, W. (1998). Metric entropy of integration operator and small ball probabilities of the Brownian sheet. *C. R. Acad. Sci. Paris* **326**, 347-352.

[17] EDMUNDS, D. E., EVANS, W. D. and HARRIS, D. J. (1988). Approximation numbers of certain Volterra integral operators. *J. London Math. Soc.* **37**, 471–489.

[18] EDMUNDS, D. E., EVANS, W. D. and HARRIS, D. J. (1997). Two–sided estimates of the approximation numbers of certain Volterra integral operators. *Studia Math.* **124**, 59–80.

[19] EDMUNDS, D. E. and TRIEBEL, H. (1996). Function Spaces, Entropy Numbers and Differential Operators *Cambridge Univ. Press.* Cambridge.

[20] EVANS, W. D., HARRIS, D. J. and LANG, J. (1998). Two–sided estimates for the approximation numbers of Hardy–type operators in L_∞ and L_1. *Stud. Math.* **130**, 171–192.

[21] FELLER, W. (1966). An Introduction to Probability Theory and its Applications. *Wiley* New York.

[22] FERNIQUE, X. (1997). Fonctions aléatoires gaussiennes vecteurs aléatoires gaussiens. *Les Publications CRM* Montreal.

[23] GLUSKIN, E. D. (1983). Norms of random matrices and diameters of finite–dimensional sets (Russian). *Mat. Sb.* **120**, 180–189.
[24] ITÔ, K. and MCKEAN, H.P. (1965). Diffusion Processes and Their Sample Paths. *Springer Verlag* Berlin.
[25] KÖNIG, H.(1989). Eigenvalue Distribution of Compact Operators. *Birkhäuser* Boston.
[26] KUELBS, J. and LI, W. V. (1993). Metric entropy and the small ball problem for Gaussian measures. *J. Funct. Anal.* **116**, 133-157.
[27] LEDOUX, M. (1996). Isoperimetry and Gaussian analysis. *Lectures on Probability Theory and Statistics. Lecture Notes in Math.* **1648**, 165-296.
[28] LEDOUX, M. and TALAGRAND, M. (1991). Probability in Banach Spaces. *Springer Verlag* Berlin.
[29] LI, W. V. (1999). Small deviations for Gaussian Markov processes under the sup–norm. *J. Theor. Probab.* **12**, 971–984.
[30] LI, W. V. (1999). Small ball estimates for Gaussian Markov processes under L_p–norm. To appear *Stoch. Proc. Appl.*.
[31] LI, W. V. (1999). A Gaussian correlation inequality and its applications to small ball probabilities. *Electronic Commun. in Probability* **12**, 111–118.
[32] LI, W. V. and LINDE, W. (1999). Approximation, metric entropy and small ball estimates for Gaussian measures. *Ann. Probab.* **27**, 1556–1578.
[33] LI, W. V. and SHAO, Q.–M. (1999). Gaussian processes: Inequalities, small ball probabilities and applications. To appear in *Stochastic Processes: Theory and Methods. Handbook of Statistics* **19**.
[34] LIFSHITS, M. A. (1995). Gaussian Random Functions. *Kluwer* Dordrecht.
[35] LIFSHITS, M. A. (1999). Asymptotic behavior of small ball probabilities. Probab. Theory and Math. Statist. Proc. VII International Vilnius Conference. Vilnius, *VSP/TEV*, 453–468.
[36] LINDE, R. (1986). s–numbers of diagonal operators and Besov embedding. *Rend. Circ. Mat. Palermo* **2**, 83–110.
[37] MAZ'JA, V. G. (1985). Sobolev Spaces. *Springer Verlag* Berlin.
[38] PIETSCH, A. (1978). Operator Ideals. *Verlag der Wissenschaften* Berlin.
[39] PIETSCH, A. (1987). Eigenvalues and s–Numbers. *Cambridge Univ. Press.* Cambridge.
[40] PISIER, G. (1989). The Volume of Convex Bodies and Banach Space Geometry. *Cambridge Univ. Press.* Cambridge.
[41] STEPANOV, V. D. (1994). Weighted norm inequalities for integral operators and related topics. Nonlinear analysis, function spaces and applications. Proceedings of the spring school held in Prague 1994. Prague: Prometheus Publishing House, 139-175.
[42] SUDAKOV, V. N. (1969). Gaussian measures, Cauchy measures and ϵ-entropy. *Soviet Math. Dokl.* **10**, 310-313.
[43] TALAGRAND, M. (1994). The small ball problem for the Brownian sheet. *Ann. Probab.* **22**, 1331-1354.
[44] TOMCZAK–JAEGERMANN, N. (1987). Dualité des nombres d'entropie pour des opérateurs á valeurs dans un espace de Hilbert. *C.R. Acad. Sci. Paris* **305**, 299-301.

Editorial Information

To be published in the *Memoirs*, a paper must be correct, new, nontrivial, and significant. Further, it must be well written and of interest to a substantial number of mathematicians. Piecemeal results, such as an inconclusive step toward an unproved major theorem or a minor variation on a known result, are in general not acceptable for publication. Papers appearing in *Memoirs* are generally longer than those appearing in *Transactions*, which shares the same editorial committee.

As of January 31, 2002, the backlog for this journal was approximately 5 volumes. This estimate is the result of dividing the number of manuscripts for this journal in the Providence office that have not yet gone to the printer on the above date by the average number of monographs per volume over the previous twelve months, reduced by the number of volumes published in four months (the time necessary for preparing a volume for the printer). (There are 6 volumes per year, each containing at least 4 numbers.)

A Consent to Publish and Copyright Agreement is required before a paper will be published in the *Memoirs*. After a paper is accepted for publication, the Providence office will send a Consent to Publish and Copyright Agreement to all authors of the paper. By submitting a paper to the *Memoirs*, authors certify that the results have not been submitted to nor are they under consideration for publication by another journal, conference proceedings, or similar publication.

Information for Authors

Memoirs are printed from camera copy fully prepared by the author. This means that the finished book will look exactly like the copy submitted.

The paper must contain a *descriptive title* and an *abstract* that summarizes the article in language suitable for workers in the general field (algebra, analysis, etc.). The *descriptive title* should be short, but informative; useless or vague phrases such as "some remarks about" or "concerning" should be avoided. The *abstract* should be at least one complete sentence, and at most 300 words. Included with the footnotes to the paper should be the 2000 *Mathematics Subject Classification* representing the primary and secondary subjects of the article. The classifications are accessible from www.ams.org/msc/. The list of classifications is also available in print starting with the 1999 annual index of *Mathematical Reviews*. The Mathematics Subject Classification footnote may be followed by a list of *key words and phrases* describing the subject matter of the article and taken from it. Journal abbreviations used in bibliographies are listed in the latest *Mathematical Reviews* annual index. The series abbreviations are also accessible from www.ams.org/publications/. To help in preparing and verifying references, the AMS offers MR Lookup, a Reference Tool for Linking, at www.ams.org/mrlookup/. When the manuscript is submitted, authors should supply the editor with electronic addresses if available. These will be printed after the postal address at the end of the article.

Electronically prepared manuscripts. The AMS encourages electronically prepared manuscripts, with a strong preference for $\mathcal{AMS}$-LaTeX. To this end, the Society has prepared $\mathcal{AMS}$-LaTeX author packages for each AMS publication. Author packages include instructions for preparing electronic manuscripts, the *AMS Author Handbook*, samples, and a style file that generates the particular design specifications of that publication series. Though $\mathcal{AMS}$-LaTeX is the highly preferred format of TeX, author packages are also available in $\mathcal{AMS}$-TeX.

Authors may retrieve an author package from e-MATH starting from www.ams.org/tex/ or via FTP to ftp.ams.org (login as anonymous, enter username as password, and type cd pub/author-info). The *AMS Author Handbook* and the *Instruction Manual* are available in PDF format following the author packages link from www.ams.org/tex/. The author package can be obtained free of charge by sending email to pub@ams.org (Internet) or from the Publication Division, American Mathematical Society, P.O. Box 6248, Providence, RI 02940-6248. When requesting an author package, please specify $\mathcal{AMS}$-LaTeX or $\mathcal{AMS}$-TeX, Macintosh or IBM (3.5) format, and the publication in which your paper will appear. Please be sure to include your complete mailing address.

Sending electronic files. After acceptance, the source file(s) should be sent to the Providence office (this includes any TeX source file, any graphics files, and the DVI or PostScript file).

Before sending the source file, be sure you have proofread your paper carefully. The files you send must be the EXACT files used to generate the proof copy that was accepted for publication. For all publications, authors are required to send a printed copy of their paper, which exactly matches the copy approved for publication, along with any graphics that will appear in the paper.

TeX files may be submitted by email, FTP, or on diskette. The DVI file(s) and PostScript files should be submitted only by FTP or on diskette unless they are encoded properly to submit through email. (DVI files are binary and PostScript files tend to be very large.)

Electronically prepared manuscripts can be sent via email to pub-submit@ams.org (Internet). The subject line of the message should include the publication code to identify it as a Memoir. TeX source files, DVI files, and PostScript files can be transferred over the Internet by FTP to the Internet node e-math.ams.org (130.44.1.100).

Electronic graphics. Comprehensive instructions on preparing graphics are available at www.ams.org/jourhtml/graphics.html. A few of the major requirements are given here.

Submit files for graphics as EPS (Encapsulated PostScript) files. This includes graphics originated via a graphics application as well as scanned photographs or other computer-generated images. If this is not possible, TIFF files are acceptable as long as they can be opened in Adobe Photoshop or Illustrator. No matter what method was used to produce the graphic, it is necessary to provide a paper copy to the AMS.

Authors using graphics packages for the creation of electronic art should also avoid the use of any lines thinner than 0.5 points in width. Many graphics packages allow the user to specify a "hairline" for a very thin line. Hairlines often look acceptable when proofed on a typical laser printer. However, when produced on a high-resolution laser imagesetter, hairlines become nearly invisible and will be lost entirely in the final printing process.

Screens should be set to values between 15% and 85%. Screens which fall outside of this range are too light or too dark to print correctly. Variations of screens within a graphic should be no less than 10%.

Inquiries. Any inquiries concerning a paper that has been accepted for publication should be sent directly to the Electronic Prepress Department, American Mathematical Society, P. O. Box 6248, Providence, RI 02940-6248.

Editors

This journal is designed particularly for long research papers, normally at least 80 pages in length, and groups of cognate papers in pure and applied mathematics. Papers intended for publication in the *Memoirs* should be addressed to one of the following editors. In principle the Memoirs welcomes electronic submissions, and some of the editors, those whose names appear below with an asterisk (*), have indicated that they prefer them. However, editors reserve the right to request hard copies after papers have been submitted electronically. Authors are advised to make preliminary email inquiries to editors about whether they are likely to be able to handle submissions in a particular electronic form.

Algebra to KAREN E. SMITH, Department of Mathematics, University of Michigan, 525 University, Suite 2832, Ann Arbor, MI 48109-1109; email: `kesmith@lsa.umich.edu`

Algebraic geometry and commutative algebra to LAWRENCE EIN, Department of Mathematics, University of Illinois, 851 S. Morgan (M/C 249), Chicago, IL 60607-7045; email: `ein@uic.edu`

Algebraic topology and cohomology of groups to STEWART PRIDDY, Department of Mathematics, Northwestern University, 2033 Sheridan Road, Evanston, IL 60208-2730; email: `priddy@math.nwu.edu`

Combinatorics and Lie theory to SERGEY FOMIN, Department of Mathematics, University of Michigan, Ann Arbor, Michigan 48109-1109; email: `fomin@math.lsa.umich.edu`

Complex analysis and complex geometry to DUONG H. PHONG, Department of Mathematics, Columbia University, 2990 Broadway, New York, NY 10027-0029; email: `phong@math.columbia.edu`

*****Differential geometry and global analysis** to LISA C. JEFFREY, Department of Mathematics, University of Toronto, 100 St. George St., Toronto, ON Canada M5S 3G3; email: `jeffrey@math.toronto.edu`

Dynamical systems and ergodic theory to ROBERT F. WILLIAMS, Department of Mathematics, University of Texas, Austin, Texas 78712-1082; email: `bob@math.utexas.edu`

Functional analysis and operator algebras to DAN VOICULESCU, Department of Mathematics, University of California, Berkeley, 970 Evans Hall, Floor 9, Berkeley, CA 94720-0001; email: `dvv@math.berkeley.edu`

Geometric topology, knot theory and hyperbolic geometry to ABIGAIL A. THOMPSON, Department of Mathematics, University of California, Davis, Davis, CA 95616-5224; email: `thompson@math.ucdavis.edu`

Harmonic analysis, representation theory, and Lie theory to ROBERT J. STANTON, Department of Mathematics, The Ohio State University, 231 West 18th Avenue, Columbus, OH 43210-1174; email: `stanton@math.ohio-state.edu`

*****Logic** to THEODORE SLAMAN, Department of Mathematics, University of California, Berkeley, CA 94720-3840; email: `slaman@math.berkeley.edu`

Number theory to HAROLD G. DIAMOND, Department of Mathematics, University of Illinois, 1409 W. Green St., Urbana, IL 61801-2917; email: `diamond@math.uiuc.edu`

*****Ordinary differential equations, partial differential equations, and applied mathematics** to PETER W. BATES, Department of Mathematics, Michigan State University, East Lansing, MI 48824-1027; email: `bates@math.msu.edu`

*****Probability and statistics** to KRZYSZTOF BURDZY, Department of Mathematics, University of Washington, Box 354350, Seattle, Washington 98195-4350; email: `burdzy@math.washington.edu`

*****Real and harmonic analysis and geometric partial differential equations** to WILLIAM BECKNER, Department of Mathematics, University of Texas, Austin, TX 78712-1082; email: `beckner@math.utexas.edu`

All other communications to the editors should be addressed to the Managing Editor, WILLIAM BECKNER, Department of Mathematics, University of Texas, Austin, TX 78712-1082; email: `beckner@math.utexas.edu`.

Selected Titles in This Series

(Continued from the front of this publication)

715 **W. N. Everitt and L. Markus,** Multi-interval linear ordinary boundary value problems and complex symplectic algebra, 2001

714 **Earl Berkson, Jean Bourgain, and Aleksander Pełczynski,** Canonical Sobolev projections of weak type $(1,1)$, 2001

713 **Dorina Mitrea, Marius Mitrea, and Michael Taylor,** Layer potentials, the Hodge Laplacian, and global boundary problems in nonsmooth Riemannian manifolds, 2001

712 **Raúl E. Curto and Woo Young Lee,** Joint hyponormality of Toeplitz pairs, 2001

711 **V. G. Kac, C. Martinez, and E. Zelmanov,** Graded simple Jordan superalgebras of growth one, 2001

710 **Brian Marcus and Selim Tuncel,** Resolving Markov chains onto Bernoulli shifts via positive polynomials, 2001

709 **B. V. Rajarama Bhat,** Cocylces of CCR flows, 2001

708 **William M. Kantor and Ákos Seress,** Black box classical groups, 2001

707 **Henning Krause,** The spectrum of a module category, 2001

706 **Jonathan Brundan, Richard Dipper, and Alexander Kleshchev,** Quantum Linear groups and representations of $GL_n(\mathbb{F}_q)$, 2001

705 **I. Moerdijk and J. J. C. Vermeulen,** Proper maps of toposes, 2000

704 **Jeff Hooper, Victor Snaith, and Min van Tran,** The second Chinburg conjecture for quaternion fields, 2000

703 **Erik Guentner, Nigel Higson, and Jody Trout,** Equivariant E-theory for C^*-algebras, 2000

702 **Ilijas Farah,** Analytic guotients: Theory of liftings for quotients over analytic ideals on the integers, 2000

701 **Paul Selick and Jie Wu,** On natural coalgebra decompositions of tensor algebras and loop suspensions, 2000

700 **Vicente Cortés,** A new construction of homogeneous quaternionic manifolds and related geometric structures, 2000

699 **Alexander Fel'shtyn,** Dynamical zeta functions, Nielsen theory and Reidemeister torsion, 2000

698 **Andrew R. Kustin,** Complexes associated to two vectors and a rectangular matrix, 2000

697 **Deguang Han and David R. Larson,** Frames, bases and group representations, 2000

696 **Donald J. Estep, Mats G. Larson, and Roy D. Williams,** Estimating the error of numerical solutions of systems of reaction-diffusion equations, 2000

695 **Vitaly Bergelson and Randall McCutcheon,** An ergodic IP polynomial Szemerédi theorem, 2000

694 **Alberto Bressan, Graziano Crasta, and Benedetto Piccoli,** Well-posedness of the Cauchy problem for $n \times n$ systems of conservation laws, 2000

693 **Doug Pickrell,** Invariant measures for unitary groups associated to Kac-Moody Lie algebras, 2000

692 **Mara D. Neusel,** Inverse invariant theory and Steenrod operations, 2000

691 **Bruce Hughes and Stratos Prassidis,** Control and relaxation over the circle, 2000

690 **Robert Rumely, Chi Fong Lau, and Robert Varley,** Existence of the sectional capacity, 2000

689 **M. A. Dickmann and F. Miraglia,** Special groups: Boolean-theoretic methods in the theory of quadratic forms, 2000

688 **Piotr Hajłasz and Pekka Koskela,** Sobolev met Poincaré, 2000

For a complete list of titles in this series, visit the
AMS Bookstore at **www.ams.org/bookstore/**.